ENERGY TRANSFORMATIONS IN LIVING MATTER

A SURVEY

BY

H. A. KREBS AND H. L. KORNBERG

MEDICAL RESEARCH COUNCIL
UNIT FOR RESEARCH IN CELL METABOLISM,
DEPARTMENT OF BIOCHEMISTRY,
UNIVERSITY OF OXFORD

WITH AN APPENDIX BY
K. BURTON

WITH 21 FIGURES

SPRINGER-VERLAG
BERLIN · GÖTTINGEN · HEIDELBERG
1957

SONDERABDRUCK AUS
ERGEBNISSE DER PHYSIOLOGIE, BIOLOGISCHEN CHEMIE
UND EXPERIMENTELLEN PHARMAKOLOGIE

NEUNUNDVIERZIGSTER BAND — 1957

ALLE RECHTE,
INSBESONDERE DAS DER ÜBERSETZUNG IN FREMDE SPRACHEN,
VORBEHALTEN
OHNE AUSDRÜCKLICHE GENEHMIGUNG DES VERLAGES
IST ES AUCH NICHT GESTATTET, DIESES BUCH ODER TEILE DARAUS
AUF PHOTOMECHANISCHEM WEGE (PHOTOKOPIE, MIKROKOPIE) ZU VERVIELFÄLTIGEN

© BY SPRINGER-VERLAG OHG. BERLIN · GÖTTINGEN · HEIDELBERG 1957
PRINTED IN GERMANY

OFFSETDRUCK VON JULIUS BELTZ · WEINHEIM/BERGSTR.

Preface

This survey was written at the invitation of the Editors of the ,,Ergebnisse der Physiologie". Its aim is to present the more recent progress in the knowledge of biological energy transformations. Since it was intended for a review journal, the reader was taken to be familiar with the fundamentals of current biochemistry, as described in the standard textbooks.

It was not the object to compile an extensive collection of facts. The survey is limited to aspects of wider interest, and the main emphasis has been on the general unifying principles which emerge from the great mass of detailed observations.

The article is reprinted in the hope that it may be useful in this form to advanced students and research workers in biochemistry and related subjects.

H. A. KREBS

H. L. KORNBERG

Table of Contents[2]

	Page
1. The Key Position of Adenosine Triphosphate	213
2. The Three Phases of Foodstuff Degradation	213
3. The Energy-Yielding Steps of Intermediary Metabolism	215
4. The Build-up of Phosphate Bond Energy	221
5. Alternative Pathways of Anaerobic Fermentation in Micro-organisms	227
6. Alternative Pathways of Glucose Oxidation	237
7. The Path of Carbon in Photosynthesis	243
8. Utilization of Energy for Chemical Syntheses	249
9. Control of Energy-Supplying Processes	262
10. A Special Feature of ATP as an Energy Store	271
11. Evolution of Energy Transforming Mechanisms	273
Appendix by K. BURTON	
Free Energy Data of Biological Interest	275
References	285

The last 15 years have witnessed great advances in the analysis of the energy transformations in living organisms. From the vast amount of detailed information which has been amassed, a picture emerges revealing some striking characteristics of the chemical organization of living matter. Although metabolic processes are both diverse and complex, the number of basic components is relatively small.

[2] The following abbreviations have been used in the present survey:
AMP = adenosine 5'-phosphate; ADP = adenosine diphosphate; ATP = adenosine triphosphate; DPN = diphosphopyridine nucleotide; $DPNH_2$ = reduced diphosphopyridine nucleotide; IDP = inosine diphosphate; ITP = inosine triphosphate; GDP = guanosine diphosphate; GTP = guanosine triphosphate; P = inorganic phosphate; CoA and CoA-SH = coenzyme A; TPN = triphosphopyridine nucleotide; $TPNH_2$ = reduced triphosphopyridine nucleotide; TPP = thiamine pyrophosphate.

It has long been established that the innumerable different proteins of living matter are all built up from the same set of about 20 amino acids. It is now evident that this economy of basic units has a counterpart in dynamic biochemistry. Most of the essential machinery for the main chemical processes of living matter is provided by a comparatively small number of enzymes and coenzymes. The same catalysts are used in the different types of fermentations, in cell respiration, and in the reactions by which the constituents of living matter are synthesized.

It is true that these basic mechanisms are supplemented by many additional enzymes. This may be taken as obvious in view of the innumerable different forms in which life presents itself. But these obvious species differences which strike the eye have tended to obscure the fact that many basic features are common to many different forms of life. It is the object of this survey to discuss these common features in the sphere of energy transformations.

1. The Key Position of Adenosine Triphosphate

The concept is now firmly established (LIPMANN 1941) that the chemical energy set free in the breakdown of foodstuffs (unless it generates heat) is transformed into a special kind of chemical energy before it is converted into other forms of energy, such as mechanical work in muscle or osmotic work in secreting glands. This special form of chemical energy is that stored in the pyrophosphate bonds of adenosinetriphosphate (ATP).

Living matter shares the requirement of a special fuel with man-made machines. In a steam engine, the energy obtained from the combustion of a variety of fuels must be converted into steam pressure before it can do mechanical work. In an internal combustion engine it is the high pressure which carries out work. An electric motor must be supplied with electrical energy in a specified form to do work.

The first major stage, then, of the energy transformations in living matter culminates in the synthesis of pyrophosphate bonds of ATP, at the expense of the free energy of the degradation of foodstuffs. The overall thermodynamic efficiency of this process is estimated at 60—70%. This is high when compared with the efficiency of man-made machines depending on the burning of a fuel as a source of energy. The greater efficiency is possible because living matter is not a heat engine, but a "chemical" engine organised in a special manner.

2. The Three Phases of Foodstuff Degradation

The release of energy by the combustion of foodstuffs in living matter may be said to proceed in three major phases (Table 1).

In phase I the large molecules of the food are broken down to small constituent units. Proteins are converted to amino-acids, carbohydrates to hexoses

Table 1. *The three main phases of energy production from foodstuffs* (KREBS 1953b)

Phase	Outline of chemical change
I	carbohydrates → hexoses proteins → about 20 amino acids fats → glycerol; fatty acids
II	hexoses → lactic acid; glycerol, several amino acids (alanine, serine, cysteine) → pyruvic acid → acetyl coenzyme A; fatty acids → acetyl coenzyme A; several amino acids (3 leucines, tyrosine, phenylalanine) → acetyl coenzyme A; several amino acids (glutamic acid, histidine, prolines, arginine) → α-ketoglutaric acid; several amino acids (aspartic acid, tyrosine, phenylalanine) → oxaloacetic acid
III	acetyl coenzyme A, α-ketoglutaric acid, oxaloacetic acid → tricarboxylic acid cycle

and fats to glycerol and fatty acids. The amounts of energy liberated in the stages of phase I are relatively small. The free energy of hydrolysis of the glucosidic bonds of starch or glycogen is about 4.3 kgcal (BURTON & KREBS 1953), of the peptide bonds of the order of 3.0 kgcal (HUFFMAN 1942) and of an ester of the order of 2.5 kgcal, per mole (see KAPLAN 1951). This means that about 0.6% of the free energy of polysaccharides and proteins, and about 0.1% of that of triglyceride fats, is liberated in phase I. These quantities are not utilised except for the generation of heat. The reactions of this phase merely prepare the foodstuffs for the energy-giving processes proper: they occur mainly in the intestinal tract and in tissues when reserve material is mobilised for energy production.

In phase II the diversity of small molecules produced in the first phase — three or more different hexoses, glycerol, about twenty amino acids and a number of fatty acids — are incompletely burned, the end-product being, apart from carbon dioxide and water, one of three substances: acetic acid in the form of acetyl coenzyme A, α-ketoglutarate or oxaloacetate. The first of these three constitutes the greater amount: two-thirds of the carbon of carbohydrate and of glycerol, all carbon atoms of the common fatty acids and approximately half the carbon skeleton of amino acids yield acetyl coenzyme A. α-Ketoglutaric acid arises from glutamic acid, histidine, arginine, citrulline, ornithine, proline and hydroxyproline; oxaloacetic acid from aspartic acid and, through malic acid, from part of the benzene ring of tyrosine and phenylalanine. The details of the pathways cannot be discussed here in full [see KREBS 1954(a), GREENBERG 1954, LANG 1952, STETTEN 1955, KNOX 1955, VOGEL 1955, COON, ROBINSON & BACHHAWAT 1955]; the main stages are outlined in Table 1. One matter of principle, however, should be emphasized: the number of steps which living matter employs in order to reduce a great variety of different substances to three basic units is astonishingly small; it could certainly not be equalled with the tools at present available to the organic chemist.

The three end-products of the second phase are metabolically closely interrelated. They take part in phase III of the energy release: the tricarboxylic acid cycle, the common "terminal" pathway of oxidation of all foodstuffs. The cycle is shown in Fig. 1; it sets out how one acetic acid equivalent is burned to carbon dioxide and water, a series of di- and tri-carboxylic acids appearing as intermediate stages.

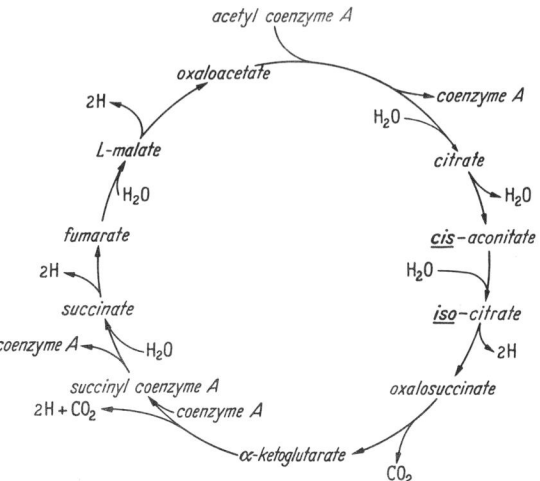

Fig. 1. *The tricarboxylic acid cycle*
Substances which enter the cycle (coenzyme A, H_2O) after the initial condensation of 1 molecule of acetyl coenzyme A and 1 molecule of oxaloacetate are written inside the cycle; substances which arise are written outside. During one turn of the cycle, one acetic acid equivalent is completely oxidized. The 4 pairs of H atoms which arise ultimately react with O_2 to form water [for further details see KREBS 1950, 1954(a)].

As each step of intermediary metabolism requires a specific enzyme, it is evident that a common pathway of oxidation results in an economy of chemical tools. Surveying the pathway of the degradation of foodstuffs as a whole, one cannot but be impressed by the relative simplicity of the arrangement, in as much as the total number of steps required to release the available energy from a multitude of different substrates is unexpectedly small.

3. The Energy-Yielding Steps of Intermediary Metabolism

The economy of stages does not, however, end here. Energy is not released at every step of intermediary metabolism. Most, though not all, of the available energy contained in the foodstuffs is liberated when the hydrogen atoms removed by dehydrogenations react with molecular oxygen to form water. Such reactions occur in phase II and in phase III. Roughly one-third of the total energy of combustion is set free in phase II and two-thirds in phase III.

Table 2. *Free energy changes in the breakdown of foodstuffs*

The figures in brackets refer to the free energy change ΔG in kilocalories per mole reactants at 25°; p_H 7.0; 0.2 atm O_2; 0.05 atm CO_2, concentrations of other reactants at unit actitivity. The pairs of H atoms removed are assumed to be oxidized to H_2O by molecular oxygen. Fatty acids are oxidized as acyl coenzyme A derivatives. The value given for the oxidative deamination of amino acids is an average figure varying a little from amino acid to amino acid. For further particulars see BURTON and KREBS (1953), KAPLAN (1951), and Tables 9—13.

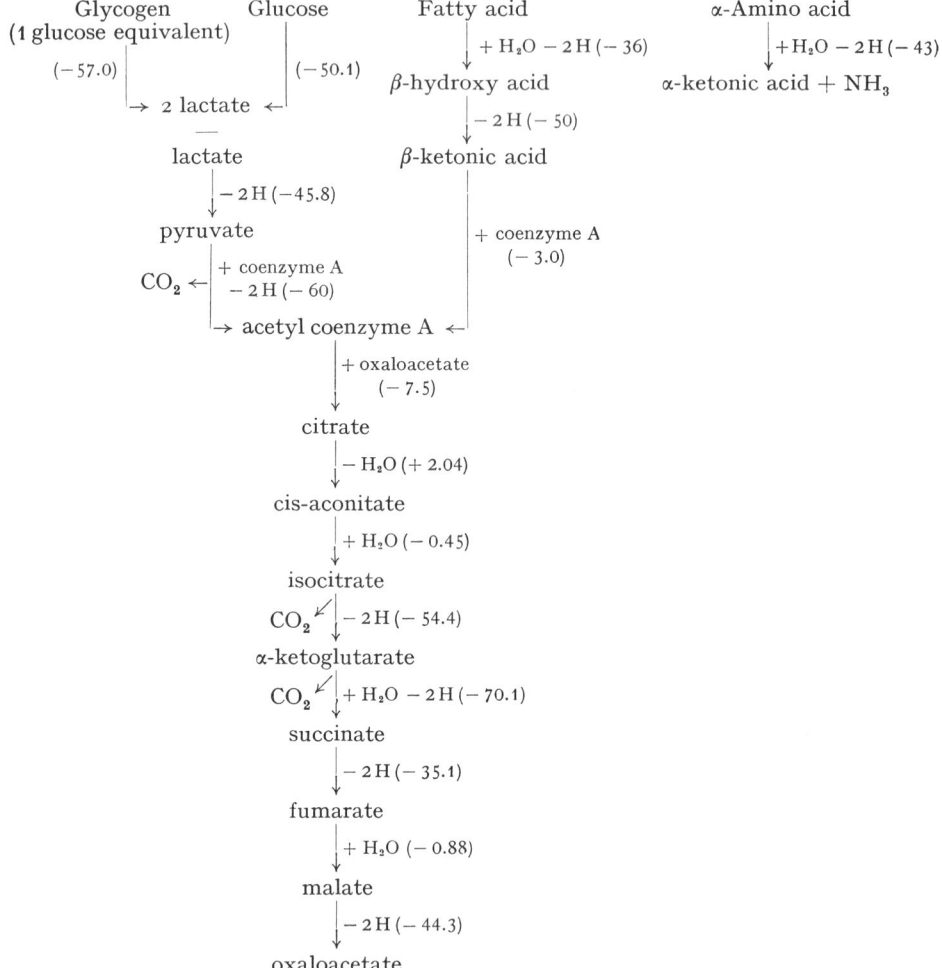

Data on the magnitude of the free energy changes have recently become available and are given in Table 2. Even with a common terminal pathway of oxidation and the shortness of the routes in phase II, the number of stages where utilizable energy is set free is still considerable (see Table 3). There are four such stages in the tricarboxylic acid cycle. In phase II there are at least twenty-nine different oxidative reactions, though more than half of these belong to one type of reaction, the oxidative deamination of amino acids.

Table 3. *Number of different oxidative stages at which energy becomes available*

"Phase" refers to the classification given in Table 1; energy is liberated when molecular oxygen is the oxidizing agent.

Phase I		0
Phase II:		
Carbohydrate → acetyl coenzyme A	2	
Fatty acids → acetyl coenzyme A	2	
Glycerol[1] → acetyl coenzyme A	1	
Amino acids:		
Oxidative deamination of amino acids[2] (about)	16	29
Oxidative decarboxylation of corresponding α-ketonic acids[3] ..	0	
Additional oxidation of phenylalanine and tyrosine leading to malic and acetoacetic acids[4]	4	
Oxidation of proline and hydroxyproline to glutamic acid ...	2	
Oxidation of arginine to glutamic acid	2	
Conversion of histidine to glutamic acid	0	
Conversion of carbon chain of leucines to acetyl coenzyme A ..	0	
Phase III		4

[1] Two more stages shared with carbohydrate.

[2] A figure lower than the total number of amino acids is given because some amino acids may share one step, e.g. (glutamic acid and histidine, serine, cystine and glycine: see BACH 1952).

[3] Assuming the mechanism to be the same as the analogous reaction of pyruvate.

[4] See WEINHOUSE and MILLINGTON (1948, 1949), SCHEPARTZ and GURIN (1949), LERNER (1949), KNOX (1955).

Their number cannot be precisely stated because the pathway of oxidation of some amino acids (e.g. tryptophan, methionine, lysine) is not yet known in every detail.

There are two oxidative stages at which carbohydrate is prepared for entry into the tricarboxylic acid cycle, namely

$$\text{triosephosphate or lactate} \rightarrow \text{pyruvate}$$
$$\text{pyruvate} \rightarrow \text{acetyl CoA}.$$

There are also two oxidative steps in the reactions which convert fatty acids to acetyl coenzyme A, namely

$$\text{fatty acid} \rightarrow \alpha\text{-}\beta\text{-unsaturated fatty acid}$$
$$\beta\text{-hydroxy fatty acid} \rightarrow \beta\text{-keto fatty acid}.$$

Glycerol derived from neutral fats, in addition to phosphorylation, requires the further special oxidative stage

$$\alpha\text{-glycerol phosphate} \rightarrow \text{triosephosphate}$$

before it joins the pathway of carbohydrate degradation. Amino acids, with a few exceptions, require at least one oxidative step each — oxidative deamination. In the cases of glutamic acid, aspartic acid, and alanine this one step is sufficient to produce an intermediate also encountered in either carbohydrate breakdown or in the tricarboxylic acid cycle. The exceptional amino acids referred to above, which do not require special oxidative steps, include histidine, which yields glutamic acid anaerobically, and cysteine and glycine, which are both convertible into serine. In many other cases additional steps are required. Details about the number of energy-releasing steps in phase II of amino acid metabolism are given in Table 3.

There are thus some thirty stages at which major parcels of energy are liberated. This is already a relatively small number, considering the variety of starting materials, but by some simple devices the number receives a further drastic reduction.

Most of the stages listed in Table 3 are complex reactions involving several distinct steps. In many cases the first of these steps is the interaction of the substrates with pyridine nucleotides, as WARBURG, CHRISTIAN and GRIESE (1935) first showed:

$$\text{substrate + pyridine nucleotide} \rightleftarrows \text{dehydrogenated substrate + reduced pyridine nucleotide} \qquad (3, 1)$$

It follows from the values of the oxido-reduction potentials of the reactants (BALL 1944, BURTON and WILSON 1952, 1953) that no appreciable amounts of energy are set free in this type of reaction, except when α-ketonic acids are oxidized, a case discussed below in more detail. The free energy of the oxidation of the substrates does not become available when the substrate is oxidized according to reaction (3, 1), but is set free when the reduced forms of the pyridine nucleotides, through the intermediation of flavoproteins and iron porphyrin enzymes, are reoxidized by molecular oxygen. The main stages of this reaction are the following (BALL 1944):

$$\text{reduced pyridine nucleotide + flavoprotein} \rightarrow \text{reduced flavoprotein +} \text{pyridine nucleotide (standard free energy change } -11 \text{ kgcal)}, \qquad (3, 2)$$

$$\text{reduced flavoprotein + 2 ferricytochrome} \rightarrow \text{flavoprotein + 2 ferrocytochrome} \text{(standard free energy change } -16 \text{ kgcal)}, \qquad (3, 3)$$

$$\text{2 ferrocytochrome} + \tfrac{1}{2} O_2 \rightarrow \text{2 ferricytochrome} + H_2O \text{(standard free energy change } -25 \text{ kgcal)}. \qquad (3, 4)$$

The over-all effect of the three reactions is the oxidation of reduced pyridine nucleotide:

$$\text{reduced pyridine nucleotide} + \tfrac{1}{2} O_2 \rightarrow \text{pyridine nucleotide} + H_2O \text{(standard free energy change } -52 \text{ kgcal)}. \qquad (3, 5)$$

This scheme of three reactions, it should be emphasised, is a provisional and simplified version of the actual events (see HERBERT 1951) because more than one iron porphyrin and more than one flavoprotein can take part in the transfer of hydrogen or electrons, these intermediate carriers probably being arranged in series. The chain of the catalysts may also include substances other than the three main types, such as the factor discovered by SLATER (1949), the chemical identity of which is unknown, as well as vitamins K and E

(MARTIUS 1954, MARTIUS and NITZ-LITZOW 1955). Moreover, as there are two pyridine nucleotides and a number of different flavoproteins and iron porphyrins, several variants of reactions (*3*, 2), (*3*, 3) and (*3*, 4) are possible. For example MARTIUS has suggested that vitamin K replaces flavoprotein in (*3*, 2) and (*3*, 3).

Reactions of the type (*3*, 2) to (*3*, 4) are common to many of the oxidations in phases II and III: to three out of the four oxidative steps of the tricarboxylic acid cycle, to the oxidation of lactate and glycerol to pyruvate, to the oxidation of α-ketonic acids to fatty acids and carbon dioxide, and also to the stages involved in the β-oxidation of fatty acids. One stage of the four dehydrogenations of the tricarboxylic acid cycle, the dehydrogenation of succinic acid, does not require a pyridine nucleotide but is catalysed by flavoproteins and iron porphyrins only. In this case only steps (*3*, 3) and (*3*, 4) of the series are shared with the other dehydrogenase systems. This is a reflection of the thermodynamic characteristics of the succinic dehydrogenase system, whose oxido-reduction potential is less negative than that of any other substrate dehydrogenase (with the possible exception of the fatty acid dehydrogenase system which is expected to be of the same order as that of succinic dehydrogenase).

Thus, with one major and possibly a few minor exceptions, all stages listed in Table 3 release energy through the same series of reactions. The major exception is the deamination of amino acids, but this exception is probably only an apparent one. One amino acid — glutamic acid — can actually interact with pyridine nucleotides directly and virtually all amino acids, according to a discovery of BRAUNSTEIN (1947) (see also CAMMARATA & COHEN 1950), can transfer their α-amino group to α-ketoglutarate by "transamination", a reversible exchange of the keto and amino groups:

$$\text{amino acid} + \alpha\text{-ketoglutarate} \rightleftarrows \text{keto acid} + \text{glutamate}. \qquad (3, 6)$$

This reaction links the deamination of all amino acids with the dehydrogenation of glutamic acid by pyridine nucleotides:

$$\text{glutamate} + \text{pyridine nucleotide} \rightleftarrows$$
$$\alpha\text{-ketoglutarate} + NH_3 + \text{reduced pyridine nucleotide}. \qquad (3, 7)$$

BRAUNSTEIN has suggested that the reactions (*3*, 6) and (*3*, 7) represent the main mechanism of deamination, but this still remains to be proved. It is an attractive hypothesis in connexion with the thesis put forward in this paper. If it holds, it would mean that no appreciable amounts of free energy are released in the transamination reaction (*3*, 6), as this reaction is readily reversible and the concentration of the reactants remains virtually constant. Nor is there a free energy change when reaction (*3*, 7) occurs under physiological conditions. The free energy of the oxidative deamination of amino acids would therefore generally be released when the reduced pyridine nucleotide undergoes oxidation through reactions (3, 2) to (3, 4), and the sixteen energy yielding reactions of amino acids listed in Table 3 would be eliminated as separate processes.

If BRAUNSTEIN's hypothesis does not hold, alternative mechanisms of oxidative deamination would have to be postulated. The introduction of such alternative mechanisms would not, however, fundamentally change the picture. Amino acid oxidases, catalysing the interaction of amino acids with oxygen, have been isolated from animal tissues. They are flavoproteins and (as indicated by their sensitivity to cyanide) are linked *in situ* to molecular oxygen by iron porphyrins. Hence, even if amino acids were oxidized through the intermediation of these enzymes, the steps where energy is released would still be represented by reactions of type (3, 3) and (3, 4).

Such evidence as is at present available favours BRAUNSTEIN's hypothesis rather than alternative oxidative mechanisms. The activity of the amino acid oxidases, which attack the more common amino acids, is weak in animal tissues, and their physiological significance has not been established. In the kidney, they may contribute to the formation of ammonia required for excretion (see KREBS 1951). Some tissues, like cardiac muscle, are devoid of amino acid oxidases yet utilise amino acids (CLARKE and WHALER 1952). The activities of the transaminases and of glutamic dehydrogenase in these tissues are sufficiently high to account for the observed rates of amino acid utilization. These facts suggest that BRAUNSTEIN's mechanism is the main pathway of amino acid oxidation: if this concept is correct, it would further ascribe a biological significance to the transaminases, enzymes the metabolic role of which is otherwise not understood.

There are a few oxidative steps in the breakdown of amino acids, for example, of tyrosine, phenylalanine, methionine and the ω-deamination of lysine and ornithine, where the enzymic mechanisms are not yet fully known. It is possible that pyridine nucleotides, flavoproteins and iron porphyrins are the catalytic agents, as has recently been shown for proline (STRECKER and MELA 1955), in which case these reactions would comply with the general pattern. But it is also feasible that the free energy of these specialised reactions is not utilised and "lost" as heat. This may be one of the factors reponsible for the rise of the basal metabolic rate when protein serves as the source of energy — the "specific dynamic action of protein". As the amino acids in question form a relatively small proportion of the protein molecule, the loss of energy would be small.

To sum up, the energy-yielding oxidative stages of Table 3 are all complex reactions, which involve a specific primary step, or series of steps — the reaction with pyridine nucleotide and, in the case of amino acids, transamination — and a set of intermediary hydrogen carriers, reacting according to (3, 2), (3, 3) and (3, 4), common to all major stages. The release of energy occurs along this common pathway and not when the substrate reacts. This arrangement reduces the total number of energy-yielding steps to six, or more correctly to six types.

4. The Build-up of Phosphate Bond Energy

Of the six energy-yielding reactions proper three occur when reduced pyridine nucleotide is oxidized by molecular oxygen through the stages (3, 2) to (3, 4). Two more occur in anaerobic glycolysis, and the sixth during the oxidative decarboxylation of α-ketonic acids. Until recently, the oxidation of succinate was thought to involve only iron porphyrin catalysts and it was assumed that major amounts of free energy are liberated when succinate is oxidized by iron porphyrins. This would represent another type of energy yielding reaction proper. However, the purification of the succinic dehydrogenase complex by KEARNEY and SINGER (1955), and by GREEN, MII and KOHOUT (1955) has shown that this system contains a flavoprotein which acts as the primary hydrogen acceptor. It is very likely, in view of the oxidoreduction potentials of flavoprotein and of the succinate/fumarate system, that there is no major release of free energy in the hydrogen transfer from succinate to flavoprotein. Thus the liberation of energy in the oxidation of succinate is by the same types of reaction as with other substrates.

The crucial feature of these reactions, from the biological point of view, is not the liberation of energy, but the coupling of energy production with energy utilisation, i.e. with the synthesis of pyrophosphate bonds of adenosine triphosphate. The utilisation of the free energy of a chemical process requires a coupling mechanism specifically adjusted to the circumstances of the energy-yielding steps. By the reduction of the number of such steps to a minimum, living matter reduces the need for specific mechanisms and thereby effects a striking economy.

The chemical mechanism by which pyrophosphate bonds of ATP are generated is known for only two of the six energy-giving reactions — those of anaerobic glycolysis. For another, the oxidative decarboxylation of α-ketoglutarate and possible other α-ketonic acids, it is known in principle, but some uncertainty of detail remains. It is completely unknown for the remaining three reactions, namely (3, 2), (3, 3) and (3, 4).

Following a suggestion of LIPMANN (1941, 1946) the pyrophosphate bonds of adenosine triphosphate are commonly referred to as "energy rich" bonds. This is a convenient term (see GILLESPIE, MAW and VERNON 1953, KALCKAR 1954) from the biological point of view (although it does not tally with the use of the term "bond energy" in physical chemistry, where it refers to the energy required to disrupt a bond between two atoms). It expresses the fact that the hydrolysis of this bond releases a much larger amount of free energy (10 to

12 kgcal per mole) than the usual hydrolytic reactions occurring in living matter, which release 2 to 4 kgcal per mole. The energy stored in pyrophosphate bonds is often referred to, again according to LIPMANN, as "phosphate bond energy".

The generation of phosphate bond energy in anaerobic glycolysis is shown in Figs. 2 and 3. The first pyrophosphate bond of ATP is formed when glycer-

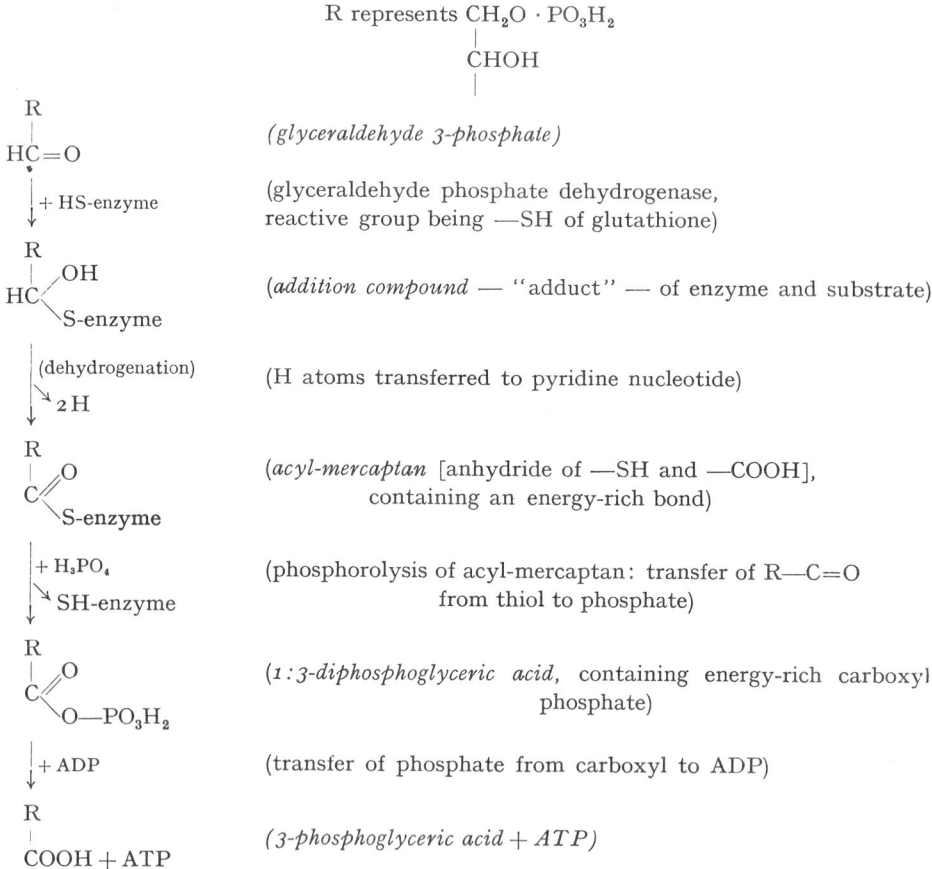

Fig. 2. *Formation of energy-rich phosphate bonds in anaerobic glycolysis: Case I*

According to WARBURG & CHRISTIAN (1939), BÜCHER (1947), RACKER & KRIMSKY (1952) and RACKER (1954)(b), energy-rich phosphate is formed during the conversion of glyceraldehyde 3-phosphate into 3-phosphoglyceric acid (see also VELICK 1954; BOYER & SEGAL 1954).

aldehyde 3-phosphate is oxidized to 3-phosphoglyceric acid. The initial step in this reaction (see Fig. 2) is taken to be the formation of an addition compound — "adduct" between the aldehyde group of the substrate and an SH group of glyceraldehyde phosphate dehydrogenase, this SH group being present in the enzyme molecule in the form of glutathione. The "adduct" undergoes a dehydrogenation, in which pyridine nucleotide bound to the enzyme acts as the hydrogen acceptor. The result is the formation of an acyl-mercaptan.

Being an acid anhydride (anhydride of —SH and —COOH), this represents an energy-rich bond. The acyl-mercaptan undergoes phosphorolysis causing a transfer of the acyl group from the sulphur atom to the phosphate group. The product is 1:3-diphosphoglyceric acid, which contains an energy-rich carboxyl phosphate. Finally, phosphate is transferred from the carboxyl to ADP, and ATP arises. These transfer reactions are readily reversible, which indicates that no major free energy change is involved.

A second molecule of ATP is formed from ADP in the later stages of anaerobic glycolysis. An energy-rich phosphate bond arises when 2-phospho-

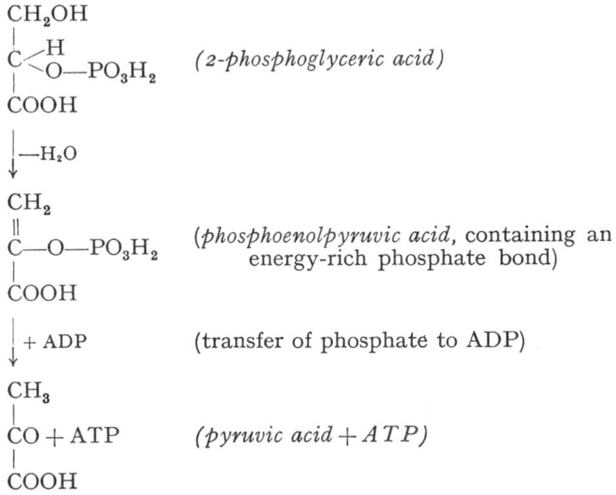

$$\begin{array}{l} CH_2OH \\ \diagup H \\ C\diagdown O-PO_3H_2 \\ \\ COOH \end{array}$$ (*2-phosphoglyceric acid*)

$$\downarrow -H_2O$$

$$\begin{array}{l} CH_2 \\ \| \\ C-O-PO_3H_2 \\ | \\ COOH \end{array}$$ (*phosphoenolpyruvic acid*, containing an energy-rich phosphate bond)

$$\downarrow + ADP$$ (transfer of phosphate to ADP)

$$\begin{array}{l} CH_3 \\ | \\ CO + ATP \\ | \\ COOH \end{array}$$ (*pyruvic acid + ATP*)

Fig. 3. *Formation of energy-rich phosphate bonds in anaerobic glycolysis: Case II*
Energy-rich phosphate is formed during conversion of 2-phosphoglyceric acid into phosphoenolpyruvic acid.

glyceric acid loses the elements of water and forms phosphoenol pyruvic acid (see Fig. 3). The subsequent cleavage to pyruvic acid is coupled with the transfer of the phosphate bond to ADP, thus generating a molecule of ATP.

The third case (later referred to as Case III) where information on the mechanism of ATP synthesis is at hand, is the phosphorylation of ADP coupled with the dehydrogenation of α-ketoglutarate. The following three reactions have been established as components of this mechanism.

α-ketoglutarate + coenzyme A + DPN $\rightleftarrows$ succinyl coenzyme A + CO_2 + $DPNH_2$ (*4*, 1)

succinyl coenzyme A + GDP + P $\rightleftarrows$ coenzyme A + GTP + succinate   (*4*, 2)

GTP + ADP $\rightleftarrows$ GDP + ATP   (*4*, 3)

Sum: α-ketoglutarate + ADP + P + DPN → succinate + CO_2 + ATP + $DPNH_2$

Reactions (*4*, 1) and (*4*, 2) are themselves the sum of a number of different steps, in which the co-factors thiamine pyrophosphate (TPP) and α-lipoic acid (6,8-dithio-n-octanoic acid) participate:

$$\begin{array}{c} S-CH{\diagup}^{(CH_2)_4 \cdot COOH}_{CH_2} \\ | \\ S-CH_2 \end{array}$$

α-lipoic acid

(REED, DE BUSK, GUNSALUS and HORNBERGER 1951; BULLOCK, BROCKMAN, PATTERSON, PIERCE and STOKSTAD 1952.)

A tentative reaction mechanism for the overall reaction (4, 1) has been formulated by GUNSALUS (1954) (a, b) as follows [see also REED 1953, GREEN 1954 (a)]:

$$R \cdot \overset{O}{\underset{\|}{C}} \cdot \overset{O}{\underset{\|}{C}} \cdot OH + TPP \to [R \cdot \overset{O}{\underset{\|}{C}} \cdot H]\,TPP + CO_2 \qquad (4, 4)$$
(where R = —CH$_2$·CH$_2$·COOH)

$$[R \cdot \overset{O}{\underset{\|}{C}} \cdot H]\,TPP + S-CH{\diagup}^{R'}_{CH_2} \longrightarrow R \cdot \overset{O}{\underset{\|}{C}} \cdot S-CH{\diagup}^{R'}_{CH_2} + TPP \qquad (4, 5)$$
$$\qquad\qquad\qquad\quad \underset{S-CH_2}{|} \qquad\qquad\qquad\qquad HS-CH_2$$

$$R \cdot \overset{O}{\underset{\|}{C}} \cdot S-CH{\diagup}^{R'}_{CH_2} + CoA \cdot SH \longrightarrow R \cdot \overset{O}{\underset{\|}{C}} \cdot S-CoA + {\underset{HS-CH_2}{HS-CH}}{\diagup}^{R'}_{CH_2} \qquad (4, 6)$$
$$\quad HS-CH_2$$

$${\underset{HS-CH_2}{HS-CH}}{\diagup}^{R'}_{CH_2} + DPN \longrightarrow {\underset{S-CH_2}{S-CH}}{\diagup}^{R'}_{CH_2} + DPNH_2 \qquad (4, 7)$$

In this sequence the first step (4, 4) is taken to be a reaction between α-ketoglutarate and TPP in which a succinic semi-aldehyde-TPP complex is formed and CO$_2$ liberated (SANADI, LITTLEFIELD and BOCK 1952; REED 1953). The succinic semi-aldehyde-TPP complex is then thought to react with the disulphide form of α-lipoic acid (4,5) in such a manner that the aldehyde group of the TPP complex is oxidized to the corresponding carboxyl and the disulphide reduced to the dimercaptan; further that the nascent carboxyl and one of the nascent thiol groups condense to form succinyl lipoic acid, which, being an acyl mercaptan, is energy-rich. TPP is regenerated in this reaction.

By reaction (4,9) the succinyl group is transferred from the α-lipoic acid moiety to reduced coenzyme A, forming reduced α-lipoic acid and succinyl coenzyme A. In order to be able to react again in the oxidation of another molecule of aldehyde-TPP complex by (4, 5), the reduced lipoic acid has to be reoxidized to the disulphide: this is accomplished (4,7) by the action of DPN and lipoic acid dehydrogenase (HAGER and GUNSALUS 1953).

The succinyl coenzyme A reacts with GDP and inorganic phosphate (4, 2) to regenerate reduced coenzyme A and to form GTP and succinate (SANADI, GIBSON and AYENGAR 1954). GTP and ADP subsequently interact (4, 3) to form GDP and ATP (SANADI et al. 1954). It is probable that reaction (4, 2) is also a composite one, since succinyl coenzyme A can be split to succinate and coenzyme A in the presence of arsenate but in the absence of GDP (SANADI, GIBSON, AYENGAR and OUELLET 1954), and since oxygen from ^{18}O-labelled phosphate appears in one of the carboxyl groups of the succinate formed (COHN 1951). A phosphorylated succinate derivative or a phosphorylated coenzyme A may therefore be intermediates in this reaction, though neither has as yet been isolated or identified (KAUFMAN 1955, COHN 1956).

Other α-ketonic acids can react analogously to α-ketoglutarate, but in at least one case, that of pyruvate, a reaction of the type (4, 2) does not occur. In the case of pyruvate, the acetyl coenzyme A formed in a reaction similar to (4, 6) can interact with AMP to form acetyl-AMP (BERG 1955) and, subsequently, acetate and ATP:

$$\text{acetyl-S—CoA} + \text{AMP} \rightleftarrows \text{acetyl-AMP} + \text{CoA — SH}. \qquad (4, 8)$$

$$\text{acetyl-AMP} + \text{pyrophosphate} \rightleftarrows \text{acetate} + \text{ATP}. \qquad (4, 9)$$

Physiologically, reactions (4, 8) and (4, 9) seem to occur only from right to left. Their significance is not in the synthesis of energy-rich phosphate bonds, but in the "activation" of acetate. They convert acetate, and by analogous reactions other fatty acids [KORNBERG and PRICER 1953 (a)], into forms in which they can be metabolised.

A review of the three cases shows that the chemical mechanism by which energy-rich phosphate bonds are generated follows the same pattern. A series of reactions is so conducted that an ordinary "low energy bond", such as the ester bond of 2-phosphoglyceric acid, or a C—S bond of the adducts of mercaptans and aldehydes is transformed into an energy-rich bond. The first energy-rich bond to arise is not a pyrophosphate bond, but an acyl mercaptan bond in Case I (Fig. 2) and Case III (Reactions 4, 4—4, 6), and an enol-phosphate in Case II (Fig. 3). By reversible transfer reactions, the acyl mercaptans are converted into other energy-rich bonds, which eventually appear in ATP. With the exception of the enol phosphate formed in Case II the various energy-rich bonds are all acid anhydrides. In Case I, the energy-rich bond arises by the dehydrogenation of the addition compound of the substrate and the active group of the enzyme, an —SH group of glutathione. In Case II it is the elimination of the elements of water from 2-phosphoglyceric acid, in Case III a coupled oxido-reduction (dismutation) between an aldehyde and a disulphide, which creates an energy-rich bond.

Another way of looking upon the energy release from foodstuffs is the following. A series of preparatory reactions takes place in which the free energy change is small. These initial steps lead to the formation of ATP, which, by hydrolysis of a pyrophosphate bond, can release a relatively large amount of energy.

This situation is diagrammatically presented in Fig. 4 for the case of glycolysis. The height of the first column represents the total free energy change of glycolysis, i.e. that of the reaction

$$\text{glucose} + 2\,\text{ATP} \to 2\,\text{lactate} + 2\,\text{ADP} + 2\,\text{P}.$$

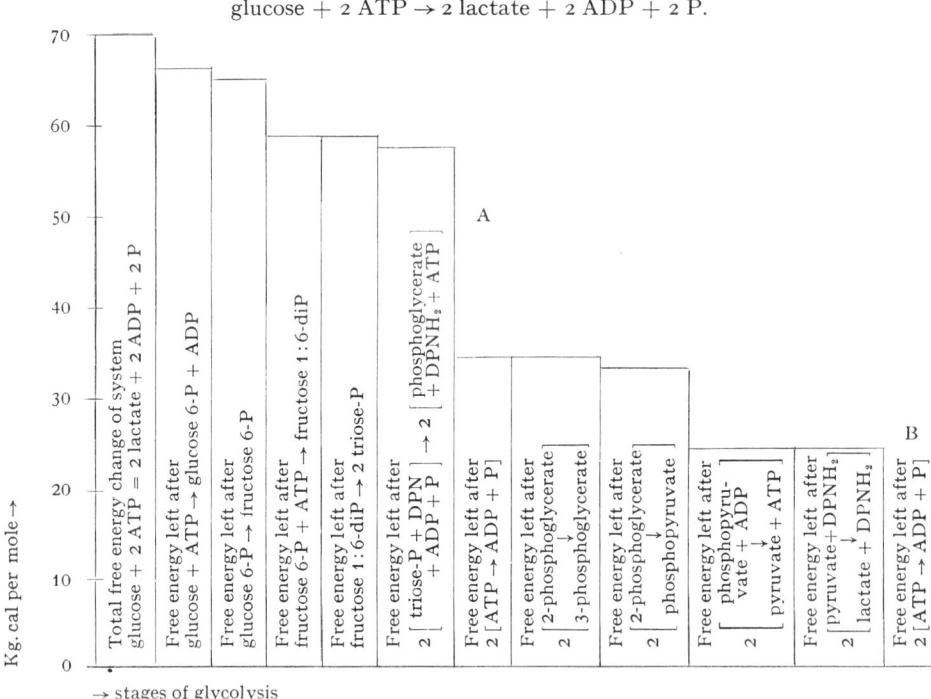

Fig. 4. *Diagrammatic presentation of the free-energy changes of glycolysis*
ATP formed in glycolysis is taken to be hydrolysed. The overall reaction is therefore
$$\text{glucose} + 2\,\text{ATP} = 2\,\text{lactate} + 2\,\text{ADP} + 2\,\text{P}.$$
Since 2 ATP molecules are synthesised during glycolysis the above equation is the sum of
$$\text{glucose} + 2\,\text{ATP} + 2\,\text{P} + 2\,\text{ADP} = 2\,\text{lactate} + 4\,\text{ADP} + 4\,\text{P}.$$
The ordinate represents the total free-energy change of this reaction (70 kgcal). As the free-energy changes depend on concentrations which are unknown, the assumption has been made that negligible free energy changes occur at those stages which are readily reversible. The magnitude of the changes at other stages is an approximate estimate (about 6 kgcal in the hexokinase, phospho-fructokinase and pyruvic kinase reactions and 11.5 kgcal for the hydrolysis of ATP).

Each column after the first represents the free energy still available for release after the various steps of the glycolytic scheme have been passed. It will be noted that the majority of steps are not accompanied by appreciable changes in free energy, and that relatively small amounts are released (and lost) at the three reactions in which ATP participates (the hexokinase, phosphofructokinase and pyruvate kinase reactions). Major quantities of energy become free, and can be utilised, only when the ATP is hydrolysed to ADP and P, at the stages marked A and B on Fig. 4. Hence the "build-up" of energy-rich phosphate bonds must not be taken to mean that energy is in any way accumulated. It means that chemical changes are so organised that a major amount of energy can be released in *one* reaction, a gradual loss in each step being avoided. The position is analogous to the utilisation of the potential energy of water placed at an elevated level. This energy cannot be transformed into other forms if the water flows to a lower level gradually, but it can drive a turbine effectively if there is a sudden drop of level.

The three reactions (Cases I, II and III) just discussed all occur anaerobically and involve substrate molecules; they are therefore often referred to as "anaerobic phosphorylations at substrate level". In contrast, the phosphorylation associated with the oxidation of reduced pyridine nucleotide does not directly involve substrates but intermediary hydrogen and electron carriers; it is usually observed when the carriers are present in catalytic quantities which are continuously reduced by the substrate and reoxidized by molecular oxygen. Because of the involvement of oxygen the mechanism is commonly referred to as "oxidative phosphorylation".

The enzymic systems of oxidative phosphorylation differ in important aspects from those of anaerobic phosphorylation at substrate level [see LEHNINGER 1955, KREBS 1954(b), SLATER 1955]. Unlike the latter, oxidative phosphorylation is inhibited by nitrophenols, halogen phenols, azide, gramicidin and other substances. At suitable concentrations, these inhibitors of phosphorylation do not inhibit the oxidations; they thus "uncouple" oxidation and phosphorylation. No similar effect is known in anaerobic phosphorylation. Furthermore, whilst the anaerobic phosphorylations occur in homogenous solutions, the mechanism responsible for oxidative phosphorylation is associated with insoluble particles of the cytoplasm, the mitochondria. Attempts to bring oxidative phosphorylation into solution have not been successful and it appears that it functions only in a heterogenous system. The mechanisms of oxidative and anaerobic phosphorylation thus seem to differ in essential points.

The free energy change of reaction $(3, 5)$ is about -52 kgcal, which is sufficient for the synthesis of at least three pyrophosphate bonds each requiring the addition of about 12 to 14 kgcal under physiological conditions. Measurements of the ratio $\frac{\text{equivalents of organic phosphate formed}}{\text{atoms of oxygen consumed}}$
show that three pyrophosphate bonds can actually be synthesised when one molecule of reduced pyridine nucleotide is oxidized (LEHNINGER 1949, 1951, 1955) and it is considered as probable, though not as certain, that one is formed in each of the three component reactions $(3, 2)$, $(3, 3)$ and $(3, 4)$ (NIELSEN and LEHNINGER 1955, BORGSTRÖM, SUDDRUTH and LEHNINGER 1955).

5. Alternative Pathways of Anaerobic Fermentation in Micro-organisms

What has been said so far refers mainly to animal tissues. The same basic mechanisms of energy release, as summarised in Tables 1 and 2, and Fig. 5, have been found in all types of animals from protozoa to mammals. They

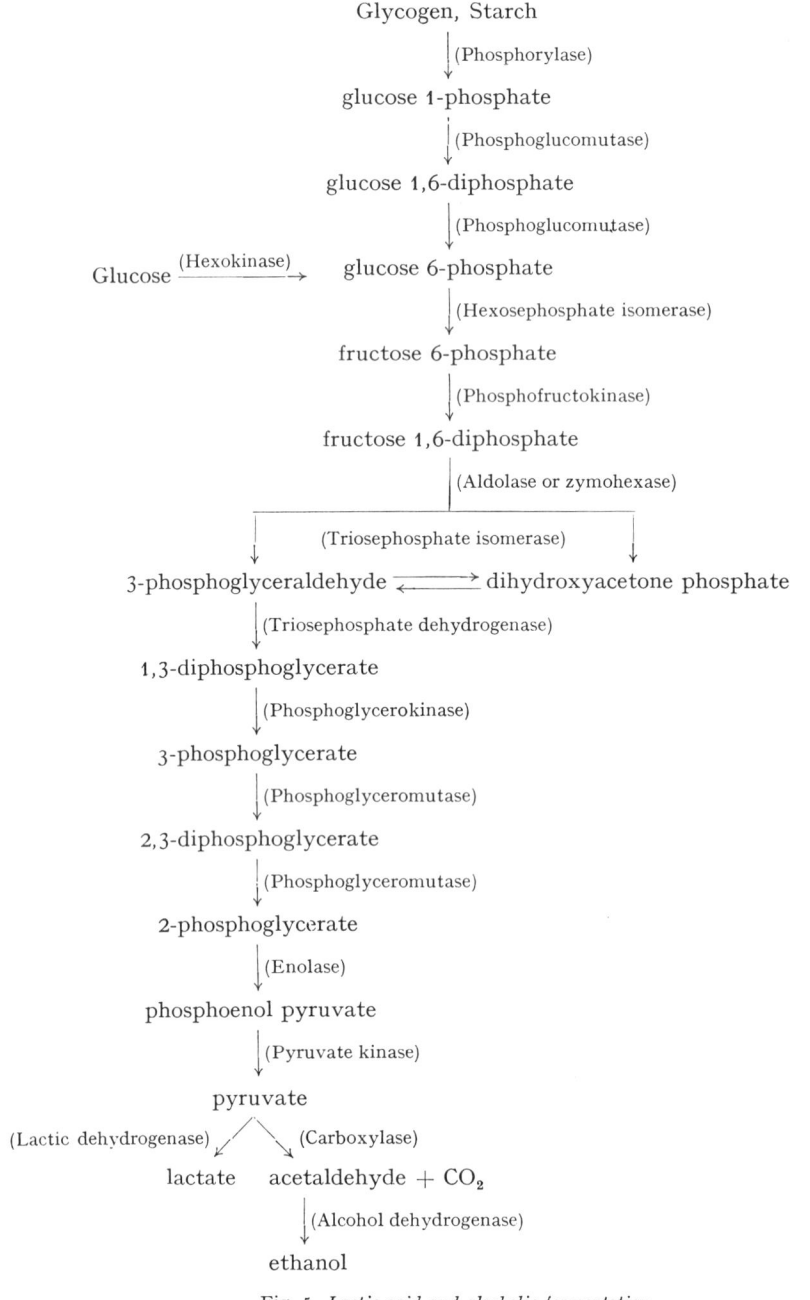

Fig. 5. *Lactic acid and alcoholic fermentation*
Intermediary changes of the carbon skeleton of carbohydrates. (The names of the enzymes responsible for each step are shown in brackets)

have also been demonstrated in a number of micro-organisms and in plants. Many micro-organisms however possess additional reactions, both aerobic and anaerobic, which release energy. Some of these are supplementary to, others take the place of, those already discussed. Furthermore, alternative modes of foodstuff degradation occur and at least one of these, the pentose phosphate cycle, appears to be present in most types of organism [see RACKER 1954(a)], though the amounts of sugars broken down by this route are usually small.

Some micro-organisms do not ferment sugar to lactic acid, but to a variety of other end-products such as formic, acetic, propionic, butyric and succinic acids, ethanol, propanol, butanol, acetylmethylcarbinol, butylene-glycol, acetone, and gaseous hydrogen. It has long been established that lactic acid and alcoholic fermentations share a common pathway. Of the twelve major intermediary stages of fermentation (Fig. 5), only the terminal ones differ in alcoholic and lactic acid fermentation, pyruvate being the stage where the branching occurs. Many observations on other forms of fermentation are consistent with the view that the intermediary stages of sugar breakdown leading to pyruvate occur also in those fermentations which yield the end-products mentioned above. Pyruvate is thus not only the branching point for the lactate and alcoholic fermentations but also for the other forms of fermentation. The formation of pyruvate from carbohydrate has been directly demonstrated in a number of species (AUBEL 1926, COOK 1930, WOOD and WERKMAN 1934, WOOD, STONE and WERKMAN 1937). It has also been shown in many cases that pyruvate can be fermented to the same products as carbohydrate [COHEN 1949, COHEN and COHEN-BAZIRE 1948, COHEN-BAZIRE et al. 1948, COHEN-BAZIRE and COHEN 1949, ROSENFELD and SIMON 1950(a), (b)]. Striking demonstrations of the existence of the lactate-producing mechanism in organisms forming other end-products are inhibitor experiments with carbon monoxide and hydrocyanic acid, and growth experiments on iron-deficient media. Carbon monoxide, or hydrocyanic acid (KEMPNER and KUBOWITZ 1933, KUBOWITZ 1934, SIMON 1947), or omission of iron from the culture medium, convert a butyric acid fermentation, or an acetone-butanol fermentation of Clostridia into a lactic acid fermentation. (The names of the various types of fermentation are derived from the predominant end-products.) These observations also indicate that the secondary mechanisms, concerned with the conversion of pyruvate into special end-products different from lactate, involve iron catalysts.

The pathway of the secondary changes of pyruvate in various types of anaerobic carbohydrate fermentation is outlined in Fig. 6. These different types (the lactic, propionic, butyric, acetic-ethanol, butanol-acetone, acetyl methyl carbinol, butylene glycol and succinic fermentations) seldom occur alone. Most of the forms mentioned can occur simultaneously with the lactic acid fermentation. The overall chemical changes in these and other fermentations are shown in Table 4.

The scheme shown in Fig. 6 has two features of special interest. The first is that the number of additional reactions is relatively small considering the

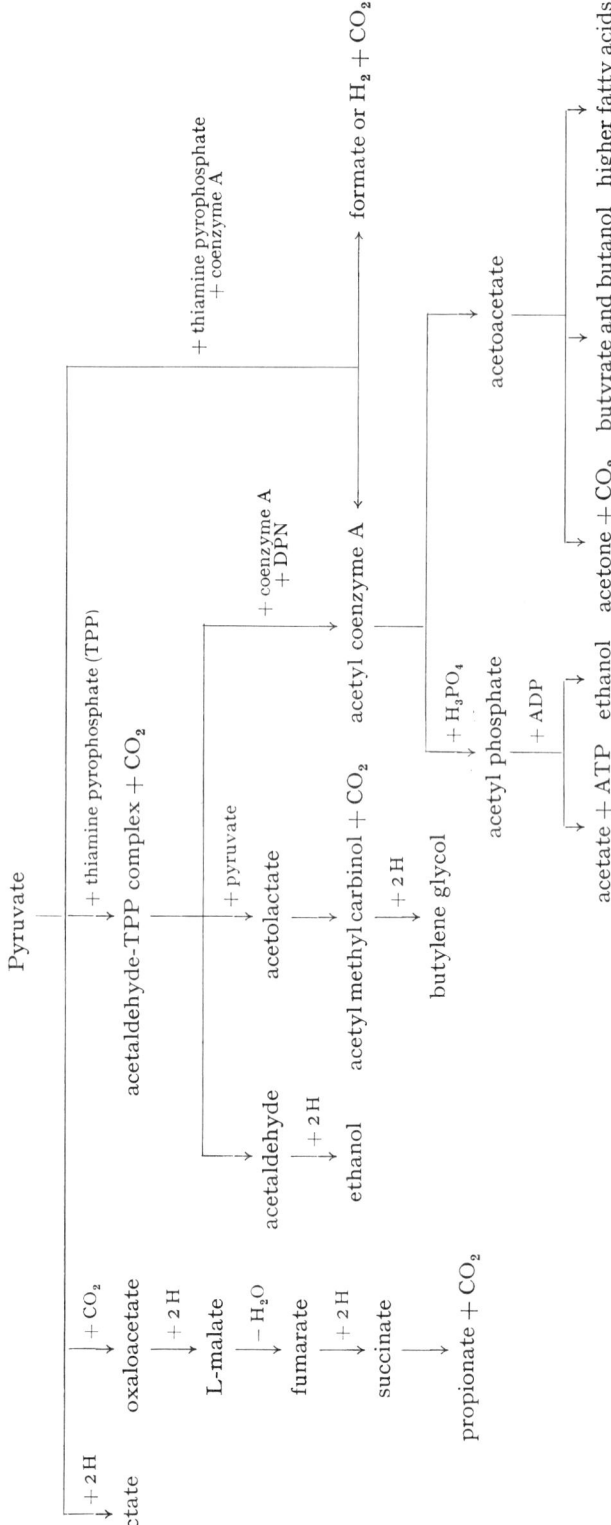

Fig. 6. *Chief reactions of pyruvate in bacterial fermentations*

Table 4. *Survey of anaerobic fermentations of carbohydrates and related substances*

(Most of the fermentations occur in combination with others. Where hydrogen atoms [H] are given as end products, these react either to form H_2, or to reduce metabolites arising from concomitant fermentations. The survey is not intended to be comprehensive. For references see STEPHENSON 1949, PORTER 1946, GUNSALUS 1947, CAMPBELL & GUNSALUS 1944, SLADE & WERKMAN 1941, BREWER & WERKMAN 1940, KREBS 1937)

Designation of fermentation	Nature of fermentation	Example of occurrence
Lactic acid	hexose → 2 lactic acid	All animal tissues; many bacteria e.g. Lactobacilli, Coliforms, Clostridia, Streptococci, Staphylococci, plant tissues
	glycerol → lactic acid + 2 H	Streptococci
Alcoholic	hexose → 2 ethanol + 2 CO_2	Yeasts, moulds, plants, many bacteria e.g. Coliforms, B. polymyxa, B. macerans
Propionic acid	1½ hexose or 3 lactic acid → 2 propionic acid + acetic acid + CO_2	Propionibacteria, Veillonella
	glycerol → propionic acid + H_2O	Propionibacteria
Succinic acid	½ hexose + CO_2 + 2 H } → succinic acid + H_2O or glycerol + CO_2	Many bacteria e.g. Coliforms, Propionibacteria
Butyric acid	hexose → butyric acid + 2 CO_2 + 2 H_2	Clostridia
Butanol-acetone	2 hexose → butanol + acetone + 5 CO_2 + 4 H_2O	Clostridium butylicum
Butylene glycol	hexose → butylene glycol + 2 CO_2 + 2 H	Aerobacter, B. subtilis, B. polymyxa, Lactobacilli
Acetylmethyl carbinol . .	hexose → acetylmethyl carbinol + 2 CO_2 + 4 H	Animal tissues, Staphylococci, Gonococci
Unnamed fermentations of intermediates . . .	2 pyruvic acid + H_2O → lactic acid + acetic acid + CO_2	Coliforms
	pyruvic acid + H_2O → acetic acid + formic acid, or CO_2 + H_2	Coliforms
	3 oxaloacetic acid + 2 H_2O → succinic acid + 2 acetic acid + 4CO_2	Coliforms
	7 fumaric acid + 4 H_2O → 6 succinic acid + 4 CO_2	Coliforms
	or 7 malic acid → 6 succinic acid + 4 CO_2 + 3 H_2O	
	3 citric acid + 2 H_2O → succinic acid + 4CO_2 + 5 acetic acid	Lactobacilli, Streptococci, Aerobacter
	2 citric acid → 2 acetylmethyl carbinol + 2 acetic acid + 2 CO_2	

variety of end-products; the second, that the occurrence of the majority of the additional reactions is by no means restricted to the organisms endowed with special fermenting powers. Several of these reactions, such as the synthesis of acetoacetate and of C_4-dicarboxylic acids, also take place in many other organisms, including higher animals, though in these not as part of the anaerobic energy-producing mechanism. This is an instance of the general experience that one and the same primary mechanism is used by living matter for a variety of different ends.

Apart from the reduction to lactate there appear to be only three major *primary* reactions of pyruvate in bacterial fermentations: (1) the carboxylation to yield oxaloacetate, or possibly (if associated with reduction) malate (WOOD 1946, UTTER & WOOD 1951) and (2) the interaction with thiamine pyrophosphate, leading to decarboxylation and to the formation of possibly an acetaldehyde thiamine pyrophosphate complex; (3) interaction with thiamine pyrophosphate and coenzyme A by a mechanism not known in detail (see WOLFE & O'KANE 1955) leading, like (2), to acetyl coenzyme A, but, unlike (2), forming either formate or $H_2 + CO_2$. These three primary reactions can be followed by various secondary reactions. Reduction of the C_4-dicarboxylic acids is probably the main pathway by which succinic acid is formed in microorganisms. Decarboxylation of succinate gives propionate, but this pathway of propionate formation does not seem to be the only one. The "acetaldehyde-thiamine pyrophosphate complex" can undergo at least three different reactions (see Fig. 6). Most widespread among these is the formation of acetyl coenzyme A, via reactions analogous to (4, 5) and (4, 6). The acetyl group of acetyl coenzyme A can undergo a variety of fermentation reactions, leading to free acetic acid, ethanol, acetone, butyric acid, butanol or higher fatty acids. Whilst many details of the pathways by which these substances are formed from acetyl coenzyme A are still unknown it is probable that again there are only two primary reactions: a condensation reaction between two molecules of acetyl coenzyme A to yield acetoacetate (STADTMAN, DOUDOROFF & LIPMANN 1951) and the transacetylation reaction, yielding acetyl phosphate (STADTMAN, NOVELLI & LIPMANN 1951). Butyric acid may be taken to be formed by the reduction of acetoacetic acid, reacting in the form of the coenzyme A derivative.

Acetyl phosphate yields free acetic acid by interaction with ADP (STADTMAN & BARKER 1950):

$$CH_3 \cdot CO \cdot OPO_3H_2 + ADP \rightarrow CH_3 \cdot CO \cdot OH + ATP. \qquad (5, 1)$$

The mechanism of formation of ethanol from acetyl phosphate by reduction of acetic acid as shown in Fig. 6, is hypothetical, but supported by some evidence. A carboxylase of the type operating in yeast fermentation is uncommon in bacteria and the mechanism of ethanol formation in bacteria is therefore likely to be different from that in yeast. Tracer experiments by Wood, Brown and Werkman (1945) have shown that bacteria can reduce acetate to ethanol, and butyrate to butanol. The possibility of an enzymic reduction of a carboxyl phosphate has been established (Bücher 1947, Baranowski 1949, von Euler, Adler and Guenther 1939) for the system 3-phospho-glyceric acid → 3-phosphoglyceraldehyde → 3-phospho-glycerol, the stages being:

$$\begin{array}{c} COOH \\ | \\ CHOH \\ | \\ CH_2OPO_3H_2 \end{array} + ATP \rightarrow \begin{array}{c} COOPO_3H_2 \\ | \\ CHOH \\ | \\ CH_2OPO_3H_2 \end{array} + ADP \qquad (5, 2)$$

3-phosphoglyceric acid   *1:3-diphosphoglyceric acid*

$$\begin{array}{c} COOPO_3H_2 \\ | \\ CHOH \\ | \\ CH_2OPO_3H_2 \end{array} + DPNH_2 \rightarrow \begin{array}{c} CHO \\ | \\ CHOH \\ | \\ CH_2OPO_3H_2 \end{array} + DPN + H_3PO_4 \qquad (5, 3)$$

1:3-diphosphoglyceric acid   *glyceraldehyde 3-phosphate*

$$\begin{array}{c} CHO \\ | \\ CHOH \\ | \\ CH_2OPO_3H_2 \end{array} \rightleftarrows \begin{array}{c} CH_2OH \\ | \\ CO \\ | \\ CH_2OPO_3H_2 \end{array} \qquad (5, 4)$$

glyceraldehyde 3-phosphate   *dihydroxyacetone phosphate*

$$\begin{array}{c} CH_2OH \\ | \\ CO \\ | \\ CH_2OPO_3H_2 \end{array} + DPNH_2 \rightarrow \begin{array}{c} CH_2OH \\ | \\ CHOH \\ | \\ CH_2OPO_3H_2 \end{array} + DPN \qquad (5, 5)$$

dihydroxyacetone phosphate   *α-glycerophosphate*

Butanol may be formed by analogous reactions from butyryl phosphate, which is an intermediate in the metabolism of Clostridium butylicum. Alternatively ethanol and its homologues might involve a reaction of the type

acetyl coenzyme A + $DPNH_2$ ⇄ acetaldehyde + coenzyme A + DPN

(Burton & Stadtman 1953).

The quantities of free energy which become available in the different forms of anaerobic fermentations are listed in Tables 5 and 6. The data are approximate because of possible errors in the basic data used for the calculation. Moreover they refer to standard concentrations and not to actual concentrations which are variable and therefore not definable. To compare the different forms of fermentation, the quantities of energy obtained from one half glucose equivalent have also been listed. Substantially greater amounts are released from the more complex fermentations shown in Table 5. On the other hand, the free energy changes accompanying the heterofermentations listed in Table 6

Table 5. *Free-energy changes of anaërobic fermentations*

The data are calculated from those given by BURTON and KREBS (1953); ΔG at p_H 7.0; standard activities

Reaction	ΔG (kgcal)	ΔG per $1/2$ glucose eq. (kgcal)
Glucose → 2 lactate⁻ + 2 H⁺	− 50.0	− 25.0
Glucose → 2 ethanol + 2 CO_2	− 62.3	− 31.2
1½ Glucose → 2 propionate⁻ + acetate⁻ + 3 H⁺ + CO_2 + H_2O	−113.2	− 37.7
3 Lactate⁻ → 2 propionate⁻ + acetate⁻ + CO_2 + H_2O	− 42.6	− 14.2
Glycerol → propionate⁻ + H⁺ + H_2O	− 36.4	− 36.4
Glycerol + CO_2 → succinate²⁻ + 2 H⁺ + H_2O	− 32.3	− 32.3
Glucose → butyrate⁻ + H⁺ + 2 CO_2 + 2 H_2	− 62.6	− 31.3
Glucose → butyrate⁻ + 3 H⁺ + 2 HCOO⁻	− 61.4	− 30.7
2 Glucose → butanol + acetone + 5 CO_2 + 4 H_2	−112.1	− 28.0

Table 6. *Free-energy changes of anaerobic heterofermentations of E. coli*

The data are calculated from those given by BURTON and KREBS (1953); ΔG at p_H 7.0; standard activities. The reaction schemes are formulated on the basis of the findings of the authors quoted [KREBS 1954 (b)]

Author	Reaction	ΔG (kgcal)	ΔG (per $1/2$ glucose eq.) (kgcal)
HARDEN (1901)	2 Glucose + H_2O → 2 lactate⁻ + acetate + ethanol + CO_2 + 2 H_2 + 3 H⁺	− 101.4	− 25.3
STOKES (1949)	5 Glucose + 2 H_2O → 2 succinate²⁻ + 4 acetate⁻ + 4 ethanol + 6 fumarate²⁻ + 14 H⁺	− 288.7	− 28.9
SCHEFFER (1928)	5 Glucose → 2 succinate²⁻ + 2 acetate⁻ + 2 ethanol + 2 CO_2 + 2 H_2 + 4 lactate⁻ + 10 H⁺	− 276.4	− 27.6

yield surprisingly little extra energy. This suggests that the significance of the secondary reactions in these organisms may lie in the synthesis of intermediates required for growth, rather than in the supply of energy. Acetyl coenzyme A as well as C_4-dicarboxylic acids can act as precursors of many cell constituents.

The great variety of additional fermentation reactions appears to include only one additional reaction, or type of reaction, which leads directly to the synthesis of ATP and thus belongs to the energy-giving reactions proper. This is the acyl-phosphokinase reaction (STERN and OCHOA 1951, STADTMAN 1952):

$$\text{acyl phosphate} + \text{ADP} \rightleftarrows \text{fatty acid} + \text{ATP}. \tag{5, 6}$$

There is one reaction of this type which has been definitely established involving acetic acid:

$$\text{acetyl phosphate} + \text{ADP} \rightleftarrows \text{acetic acid} + \text{ATP}. \qquad (5, 1)$$

This "acetate kinase" reaction, as well as the "phosphotransacetylase" reaction preceding it:

$$\text{acetyl coenzyme A} + \text{phosphate} \rightleftarrows \text{acetyl phosphate} + \text{coenzyme A} \qquad (5, 7)$$

are absent from the animal. They are probably the main source of ATP in propionic acid bacteria or Clostridia growing anaerobically in media in which lactate is the main source of carbon, and also in Clostridia depending on the "Stickland reaction" as a source of energy. All types of Stickland reaction (coupled oxido reductions between pairs of amino acids) lead to the formation of an α-ketonic acid (NISMAN 1954) which can react with coenzyme A according to the general scheme [cf. (4, 1)]:

$$R \cdot CO \cdot COOH + \text{coenzyme A} + DPN \rightarrow R \cdot CO \cdot \text{coenzyme A} + CO_2 + DPNH_2. \quad (5, 8)$$

The acyl coenzyme A formed can yield acetyl coenzyme A by a transfer reaction catalysed by "coenzyme A transphorase" (STADTMAN 1952):

$$\text{acyl coenzyme A} + \text{acetate} \rightleftarrows \text{acetyl coenzyme A} + \text{fatty acid}. \qquad (5, 9)$$

Whether acyl coenzyme A formed by reaction (5, 8) yields ATP via reactions (5, 9), (5, 7) and (5, 1), or via a phosphotransacetylase reaction analogous to (5, 7) followed by reaction (5, 6), is an open question.

Reaction (5, 1) is the only known additional anaerobic energy-giving reaction proper of *widespread* occurrence in micro-organisms and absent from the animal. However, there is no doubt that further anaerobic sources of energy are available to special micro-organisms, or under special conditions. One such reaction has recently been discovered independently by KNIVETT (1952) and by SLADE and SLAMP (1952) and studied by OGINSKY and GEHRIG (1953), KREBS, EGGLESTON and KNIVETT (1955) and JONES, SPECTOR and LIPMANN [1955 (a, b)]. This is the decomposition of citrulline to ornithine, ammonia and CO_2:

$$\underset{\text{(citrulline)}}{NH_2 \cdot CO \cdot NH \cdot CH_2 \cdot R} + H_2O \rightarrow \underset{\text{(ornithine)}}{NH_2 \cdot CH_2 \cdot R} + CO_2 + NH_3 \qquad (5, 10)$$

[where $R = -(CH_2)_2 CH(NH_2) \cdot COOH$].

This is a component reaction of the "arginine dihydrolase" reaction discovered by HILLS (1940),

$$\text{arginine} + 2\,H_2O \rightarrow \text{ornithine} + CO_2 + NH_3$$

the first stage being the reaction

$$\text{arginine} + H_2O \rightarrow \text{citrulline}.$$

Only recently has it been recognised that reaction (5, 10) is an energy-giving process. It requires the presence of inorganic phosphate and ADP and the full chemical change is

$$\text{citrulline} + \text{phosphate} + \text{ADP} \rightarrow \text{ornithine} + NH_3 + CO_2 + \text{ATP}. \qquad (5, 11)$$

This reaction has so far been resolved into two steps. In the first carbamyl phosphate ($NH_2 \cdot CO \cdot OPO_3H_2$) is formed:

$$\text{citrulline} + \text{phosphate} \rightarrow \text{ornithine} + \text{carbamyl phosphate} \qquad (5, 12)$$

ATP arises in the second step:

$$NH_2 \cdot CO \cdot OPO_3H_2 + \text{ADP} \rightleftarrows NH_3 + CO_2 + \text{ATP}. \qquad (5, 13)$$

These reactions are in effect (and possibly, though not necessarily, in mechanism) the reversal of a stage of the synthesis of urea in mammalian liver, and of arginine synthesis in living matter generally. Reaction (5, 13), in reverse, is also a step in the synthesis of carbamyl aspartic acid which serves as a precursor in the biosynthesis of the pyrimidine ring.

The reaction (5, 11) has been studied in Streptococcus faecalis and in Clostridium perfringens (SCHMIDT, LOGAN and TYTELL 1952). It is presumably in part responsible for the accelerated growth of S. faecalis on a medium rich in arginine, arginine being a ready precursor of citrulline; but since arginine or citrulline are available only in very special circumstances the scope of this energy-giving source is obviously limited. It is, however, of interest that this energy-giving reaction is probably an adaptation evolved from the reverse process, which plays a part in the synthetic reactions of growing bacteria generally.

Aerobically micro-organisms synthesise ATP by the same reactions which occur in animal tissues namely by oxidative phosphorylation [reactions (3, 3), (3, 4) and (3, 5)]. In some cases this has been directly demonstrated (HERSEY & AJL 1951, PINCHOT 1953, HYNDMAN, BURRIS and WILSON 1943, TISSIÈRES and SLATER 1955). In others it may be inferred from the presence in many bacteria of the required reactants — pyridine nucleotides, flavoproteins and iron porphyrins. The reduction of pyridine nucleotide which provides the substrate for oxidative phosphorylation can in some micro-organisms, e.g. Micrococcus lysodeicticus, Azotobacter and Aerobacter aerogenes, be effected by the tricarboxylic acid cycle and its associated reactions. However, this cycle does not seem to operate in all organisms which oxidise carbohydrate, amino acids or fat. Moreover, micro-organisms are capable of deriving energy from the oxidation of a much larger variety of substances than are animal tissues. This applies, for example, to Pseudomonads which can utilise many organic substances inert in other organisms, and to autotrophic organisms which can obtain energy from such reactions as

$$S^= + 2\,O_2 \to SO_4^=\ \text{(Thiobacillus)}$$
$$NO_2^- + \tfrac{1}{2}O_2 \to NO_3^-\ \text{(Nitrobacter)}$$
$$NH_4^+ + 1\tfrac{1}{2}O_2 \to H_2O + NO_2^- + 2\,H^+\ \text{(Nitrosomonas)}$$
$$H_2 + \tfrac{1}{2}O_2 \to H_2O\ \text{(Hydrogenomonas)}.$$

Information on the energy transformations in these organisms is still scanty. There is no doubt that ATP and pyridine nucleotides participate also here. The energy-giving reactions proper are therefore probably identical with those in oxidative phosphorylation elsewhere. What distinguishes the energy metabolism of these organisms from that of others is their ability to use a greater variety of substrates, organic and inorganic, for the reduction of pyridine nucleotide. This is borne out by the fact that adaptation of Pseudomonads to mannitol and dulcitol (SHAW 1956) or to steroids (TALALAY and DOBSON 1953, TALALAY and MARCUS 1954) represents a formation of specific dehydrogenases catalysing the reduction of DPN by these substrates.

6. Alternative Pathways of Glucose Oxidation

The most widespread mode of biological glucose degradation consists of primary fermentation and subsequent oxidation of the products of fermentation, but alternative pathways of glucose oxidation, operating without primary fission of glucose to two triose molecules, exist in many types of different organisms. Of various routes, the most common one is the "pentose phosphate cycle". The first reaction of this cycle, the "direct" oxidation of glucose 6-phosphate (i.e. direct without prior fission to triose or triosephosphate), was discovered by WARBURG, CHRISTIAN and GRIESE in 1935 in mammalian red blood cells treated with methylene blue. The primary hydrogen acceptor in this reaction was identified as triphosphopyridine nucleotide. It was also noted (WARBURG and CHRISTIAN 1936) that the 6-phosphogluconic acid, formed from glucose 6-phosphate, undergoes further oxidation if TPN is present. Per molecule of glucose 6-phosphate 3 molecules of O_2 were used and 3 molecules of CO_2 produced; they were thus led to speak of a "combustion" of glucose 6-phosphate by TPN. Thanks to the early work of DICKENS [1936, 1938(a, b)], LIPMANN (1936) and DISCHE (1938), and the more recent work of SCOTT and COHEN (1951, 1953), DISCHE (1951) and especially of HORECKER and of RACKER and their collaborators (reviewed by RACKER [1954(a)]) and GUNSALUS, HORECKER and WOOD (1955)), it is now possible to formulate a detailed scheme of the intermediary stages of the oxidation of glucose 6-phosphate. This is the pentose phosphate cycle. The main components of the cycle are eight different reactions, (6, 1) to (6, 8). In the first reaction glucose 6-phosphate is oxidized to 6-phospho-gluconolactone (WARBURG and CHRISTIAN 1936, CORI and LIPMANN 1952, BRODIE and LIPMANN 1955), which is subsequently hydrolysed by a "lactonase" to 6-phosphogluconate.

$$\begin{array}{ccc}
\text{O=C--H} & \text{O=C------} & \text{O=C--OH} \\
\text{H--C--OH} & \text{H--C--OH} & \text{H--C--OH} \\
\text{HO--C--H} \quad \xrightleftharpoons[\text{TPNH}_2]{\text{TPN}} & \text{HO--C--H} \quad \text{O} \xrightleftharpoons[-\text{H}_2\text{O}]{+\text{H}_2\text{O}} & \text{HO--C--H} \\
\text{H--C--OH} & \text{H--C--OH} & \text{H--C--OH} \\
\text{H--C--OH} & \text{H--C------} & \text{H--C--OH} \\
\text{CH}_2\text{OPO}_3\text{H}_2 & \text{CH}_2\text{OPO}_3\text{H}_2 & \text{CH}_2\text{OPO}_3\text{H}_2 \\
\textit{Glucose 6-phosphate} & \textit{6-phospho-gluconolactone} & \textit{6-phosphogluconate}
\end{array} \quad (6, 1)$$

The 6-phosphogluconate formed is oxidatively decarboxylated (6, 2) to yield ribulose 5-phosphate, whilst another molecule of TPN is reduced (WARBURG and CHRISTIAN 1937, HORECKER, SMYRNIOTIS and SEEGMILLER 1951).

$$\begin{array}{ccc}
\text{O=C--OH} & \left[\begin{array}{c}\text{O=C=OH} \\ \text{H--C--OH} \\ \text{C=O} \\ \text{H--C--OH} \\ \text{H--C--OH} \\ \text{CH}_2\text{OPO}_3\text{H}_2\end{array}\right] & \begin{array}{c}\text{CO}_2 \\ + \\ \text{CH}_2\text{OH} \\ \text{C=O} \\ \text{H--C--OH} \\ \text{H--C--OH} \\ \text{CH}_2\text{OPO}_3\text{H}_2\end{array} \\
\text{H--COOH} & & \\
\text{HO--C--H} \quad \xrightleftharpoons[\text{TPNH}_2]{\text{TPN}} & & \\
\text{H--C--OH} & & \\
\text{H--C--OH} & & \\
\text{CH}_2\text{OPO}_3\text{H}_2 & & \\
\textit{6-phosphogluconate} & \textit{Hypothetical intermediate 3-keto} & \textit{ribulose 5-phosphate} \\
& \textit{6-phosphogluconic acid} &
\end{array} \quad (6, 2)$$

Ribulose 5-phosphate undergoes two different isomerisations: one to ribose 5-phosphate (6, 3) catalysed by pentose phosphate isomerase (AXELROD and JANG 1954)

$$\begin{array}{ccc}
\text{CH}_2\text{OH} & & \text{O=CH} \\
\text{C=O} & & \text{H--C--OH} \\
\text{H--C--OH} & \rightleftharpoons & \text{H--C--OH} \\
\text{H--C--OH} & & \text{H--C--OH} \\
\text{CH}_2\text{OPO}_3\text{H}_2 & & \text{CH}_2\text{OPO}_3\text{H}_2 \\
\textit{ribulose 5-phosphate} & & \textit{ribose 5-phosphate}
\end{array} \quad (6, 3)$$

and one (6, 4) to xylulose 5-phosphate (SRERE, COOPER, KLYBAS and RACKER 1955, DICKENS and WILLIAMSON 1955)

$$\begin{array}{ccc}
\text{CH}_2\text{OH} & & \text{CH}_2\text{OH} \\
\text{C=O} & & \text{C=O} \\
\text{H--C--OH} & \rightleftharpoons & \text{HO--C--H} \\
\text{H--C--OH} & & \text{H--C--OH} \\
\text{CH}_2\text{OPO}_3\text{H}_2 & & \text{CH}_2\text{OPO}_3\text{H}_2 \\
\textit{ribulose 5-phosphate} & & \textit{xylulose 5-phosphate}
\end{array} \quad (6, 4)$$

A molecule of xylulose 5-phosphate and one of ribose 5-phosphate, produced by the reactions (6, 3) and (6, 4), interact for form sedoheptulose 7-phosphate and glyceraldehyde 3-phosphate (6, 5) (SRERE et al. 1955, HORECKER, HURWITZ and SMYRNIOTIS 1956). This reaction is catalysed by transketolase (RACKER, DE LA HABA and LEDER 1953, DE LA HABA, RACKER and LEDER 1955), an enzyme requiring thiamine pyrophosphate as co-factor. It is thought that an "active glycolaldehyde", i.e. an [enzyme-TPP—CHO · CH$_2$OH] complex similar to the [acetaldehyde-TPP] complex mentioned in (4, 4), may be an intermediate in this reaction. This may therefore be written as:

$$\begin{array}{c}CH_2OH\\|\\C=O\\|\\HOC-H\\|\\H-C-OH\\|\\CH_2OPO_3H_2\end{array} + TPP \rightarrow \left[\begin{array}{c}CH_2OH\\|\\O=C-H\end{array}\right]TPP + \begin{array}{c}O=C-H\\|\\H-C-OH\\|\\CH_2OPO_3H_2\end{array}$$

xylulose 5-phosphate   "*active glycolaldehyde*"   *glyceraldehyde 3-phosphate*

(6, 5)

$$\left[\begin{array}{c}CH_2OH\\|\\O=C-H\end{array}\right]TPP + \begin{array}{c}O=C-H\\|\\H-C-OH\\|\\H-C-OH\\|\\H-C-OH\\|\\CH_2OPO_3H_2\end{array} \longrightarrow \begin{array}{c}CH_2OH\\|\\C=O\\|\\HO-C-H\\|\\H-C-OH\\|\\H-C-OH\\|\\H-C-OH\\|\\CH_2OPO_3H_2\end{array} + TPP$$

"*active glycolaldehyde*"   *ribose 5-phosphate*   *sedoheptulose 7-phosphate*

The glyceraldehyde 3-phosphate and sedoheptulose 7-phosphate interact further in a transfer reaction under the influence of transaldolase. This enzyme has not yet been isolated in a pure state and its co-factor requirements are therefore not known. The action of this enzyme is analogous to that of transketolase, except that the moiety transferred is not an "active glycolaldehyde" but an "active dihydroxyacetone". In this reaction (6, 6), fructose 6-phosphate and erythrose 4-phosphate are formed (HORECKER and SMYRNIOTIS 1953, 1955, HORECKER, SMYRNIOTIS, HIATT and MARKS 1955, SRERE, KORNBERG and RACKER 1955)

$$\begin{array}{c}CH_2OH\\|\\C=O\\|\\HO-C-H\\|\\H-C-OH\\|\\H-C-OH\\|\\H-C-OH\\|\\CH_2OPO_3H_2\end{array} + \begin{array}{c}O=C-H\\|\\H-C-OH\\|\\CH_2OPO_3H_2\end{array} \longrightarrow \begin{array}{c}O=C-H\\|\\H-C-OH\\|\\H-C-OH\\|\\CH_2OPO_3H_2\end{array} + \begin{array}{c}CH_2OH\\|\\C=O\\|\\HO-C-H\\|\\H-C-OH\\|\\H-C-OH\\|\\CH_2OPO_3H_2\end{array}$$

(6, 6)

sedoheptulose 7-phosphate   *glyceraldehyde 3-phosphate*   *erythrose 4-phosphate*   *fructose 6-phosphate*

The erythrose 4-phosphate formed in (6, 6) undergoes a transketolase reaction (6, 7) with a molecule of xylulose 5-phosphate. This is analogous to (6, 5) and leads to fructose 6-phosphate and glyceraldehyde 3-phosphate (KORNBERG and RACKER 1955):

$$\begin{array}{c} CH_2OH \\ | \\ C=O \\ | \\ HOC-H \\ | \\ H-C-OH \\ | \\ CH_2OPO_3H_2 \end{array} + \begin{array}{c} O=C-H \\ | \\ H-C-OH \\ | \\ H-C-OH \\ | \\ CH_2OPO_3H_2 \end{array} \longrightarrow \begin{array}{c} CH_2OH \\ | \\ C=O \\ | \\ HO-C-H \\ | \\ H-C-OH \\ | \\ H-C-OH \\ | \\ CH_2OPO_3H_2 \end{array} + \begin{array}{c} O=C-H \\ | \\ H-C-OH \\ | \\ CH_2OPO_3H_2 \end{array} \quad (6, 7)$$

xylulose 5-phosphate   erythrose 4-phosphate   fructose 6-phosphate   glyceraldehyde 3-phosphate

The fructose 6-phosphate formed in (6, 6) and (6, 7) is converted to glucose 6-phosphate by reaction (6, 8), catalysed by hexosephosphate isomerase.

$$\begin{array}{c} CH_2OH \\ | \\ C=O \\ | \\ HO-C-H \\ | \\ H-C-OH \\ | \\ H-C-OH \\ | \\ CH_2OPO_3H_2 \end{array} \rightleftarrows \begin{array}{c} H-C=O \\ | \\ H-C-OH \\ | \\ HO-C-H \\ | \\ H-C-OH \\ | \\ H-C-OH \\ | \\ CH_2OPO_3H_2 \end{array} \quad (6, 8)$$

fructose 6-phosphate   glucose 6-phosphate

This reaction completes the cycle in that it leads to the (partial) regeneration of the starting material, glucose 6-phosphate. The interplay of the components of the cycle is somewhat complex. It is shown diagrammatically in Fig. 7.

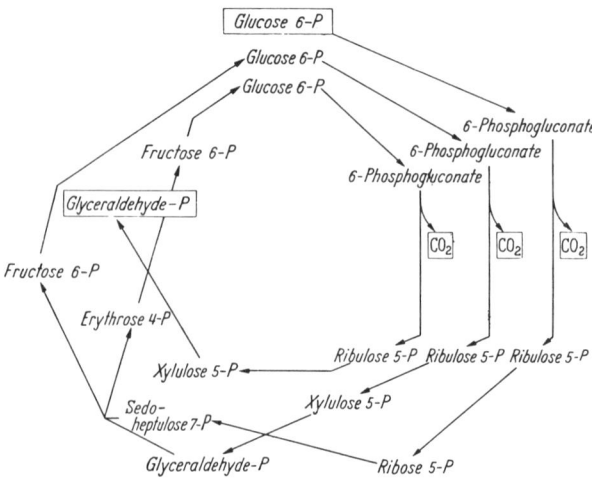

Fig. 7. *Diagram of the pentose phosphate cycle*
▭ Starting materials and end products. The crossing of arrows indicates transfer reaction. For further details see Table 7 and text

In this scheme, the reactions catalysed by transketolase and transaldolase (6, 5), (6, 6) and (6, 7) are indicated by a crossing of arrows; of the three glucose 6-phosphate molecules required for each turn of the cycle, two are regenerated. While three molecules participate in reactions (6, 1) and (6, 2), two react according to (6, 8) and only one each in the remaining reactions.

Table 7. *The component reactions of the pentose phosphate cycle and their quantitative relations*

3 glucose 6-phosphate + 3 TPN	$\xrightarrow{\text{(glucose 6-phosphate dehydrogenase)}}$	3 6-phosphogluconate + 3 TPNH$_2$ (6, 1)
3 6-phosphogluconate + 3 TPN	$\xrightarrow{\text{(6-phosphogluconate dehydrogenase)}}$	3 ribulose 5-phosphate + 3 CO$_2$ + 3 TPNH$_2$ (6, 2)
ribulose 5-phosphate	$\xrightarrow{\text{(pentose phosphate isomerase)}}$	ribose 5-phosphate (6, 3)
2 ribulose 5-phosphate	$\xrightarrow{\text{(xylulo-epimerase)}}$	2 xylulose 5-phosphate (6, 4)
ribose 5-phosphate + xylulose 5-phosphate	$\xrightarrow{\text{(transketolase)}}$	sedoheptulose 7-phosphate + glyceraldehyde 3-phosphate (6, 5)
sedoheptulose 7-phosphate + glyceraldehyde 3-phosphate	$\xrightarrow{\text{(transaldolase)}}$	fructose 6-phosphate + erythrose 4-phosphate (6, 6)
xylulose 5-phosphate + erythrose 4-phosphate	$\xrightarrow{\text{(transketolase)}}$	fructose 6-phosphate + glyceraldehyde 3-phosphate (6, 7)
2 fructose 6-phosphate	$\xrightarrow{\text{(hexose phosphate isomerase)}}$	2 glucose 6-phosphate (6, 8)
Sum: glucose 6-phosphate + 6 TPN	$\longrightarrow$	3 CO$_2$ + glyceraldehyde 3-phosphate + 6 TPNH$_2$

The net effect of one revolution of the cycle, as shown in Table 7, is therefore:

glucose 6-phosphate → glyceraldehyde 3-phosphate + 3 CO$_2$,

but the glyceraldehyde 3-phosphate thus formed does not accumulate in the organism. It can be metabolised via phosphoglycerate, phosphoenol pyruvate and pyruvate by the reactions shown in Table 2 and Fig. 1. Alternatively, if triose phosphate isomerase, aldolase, fructose 1:6-diphosphatase and hexose phosphate isomerase are present, the following sequence of reactions can occur:

glyceraldehyde 3-phosphate → dihydroxyacetone phosphate, (6, 9)

glyceraldehyde 3-phosphate + dihydroxyacetone phosphate
→ fructose 1:6-diphosphate + H$_2$O, (6, 10)

fructose 1:6-diphosphate + H$_2$O → fructose 6-phosphate + H$_3$PO$_4$, (6, 11)

fructose 6-phosphate → glucose 6-phosphate. (6, 8)

Glucose 6-phosphate would thus be formed from two molecules of glyceraldehyde 3-phosphate, and could re-enter (and be oxidized by) the pentose phosphate cycle. Reactions (6, 1) to (6, 11) repeated several times would therefore result in a complete combustion of glucose 6-phosphate. This concept, which rests on the demonstration of all the required enzymes in liver and pea root preparations (HORECKER, GIBBS, KLENOW and SMYRNIOTIS 1954; GIBBS & HORECKER 1954) is illustrated in Fig. 8.

If glucose is completely oxidized according to Fig. 8, dehydrogenations occur at only two stages (TPN being the hydrogen acceptor in both) and CO_2 is released by one reaction only. It is indeed remarkable that all six carbon atoms of glucose can be released by this one step. Again there is an extraordinary economy of chemical tools. In the glucose breakdown by fermentation followed by the tricarboxylic acid cycle there are at least 6 different reactions leading to the reduction of pyridine nucleotide, against 2 in the pentose phosphate cycle. The total number of steps by which the complete oxidation of glucose is achieved is smaller in the pentose phosphate cycle than in the breakdown by fermentation plus tricarboxylic acid cycle, but the latter route not only covers the degradation of carbohydrate but also that of fats and amino acids.

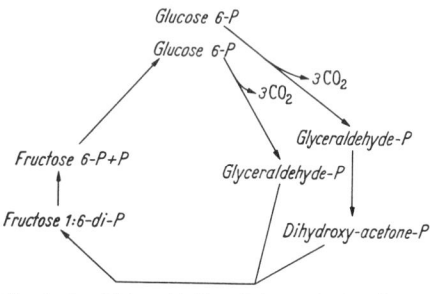

Fig. 8. *Complete oxidation of glucose 6-phosphate via the pentose phosphate cycle and additional reactions catalysed by triose phosphate isomerase, aldolase, fructose 1,6-diphosphatase and hexose phosphate isomerase.*
The first step shown in the diagram (conversion of glucose 6-phosphate to glyceraldehyde phosphate $+ 3\,CO_2$) represents the sum of the reactions shown in Fig. 7

That all carbon atoms of sugar can be released by one reaction is due to the circumstance that the cycle of Fig. 8 can effect a transfer of all carbon atoms of hexoses to position 1 from which the CO_2 is released.

The widespread occurrence in animal tissues, plants and micro-organisms of the pentose phosphate cycle has been established, but information on its quantitative significance in relation to glycolysis is still incomplete. Whilst the cycle may represent a major pathway of carbohydrate degradation in some micro-organisms, its main physiological role in most organisms is more likely to be the supply of pentose phosphate required as a constituent of nucleotides and nucleic acids, as well as the degradation of surplus pentoses in food. Erythrose 4-phosphate can also serve as a precursor in the synthesis of aromatic amino acids in micro-organisms (see chapter 8). This does not exclude that the cycle is at the same time a source of energy. Energy-giving reactions may be defined as those which lead to a synthesis of ATP. Hence all reactions in which reduced pyridine nucleotide, or reduced flavoprotein, or reduced cytochrome c, are formed, can supply utilisable energy. The reduced TPN which is formed during the two dehydrogenation steps of the cycle can probably take part in the synthesis of ATP in the same way as reduced DPN, although there is no direct evidence that TPN can directly replace DPN in reaction (*3, 5*). It is however possible that reduced TPN serves indirectly as the substrate of oxidative phosphorylation through the intermediation of DPN. Animal tissues and micro-organisms (KAPLAN, COLOWICK and NEUFELD 1952, 1953) and plants (DAVIES 1956) contain an enzyme which transfers hydrogen from TPN to DPN:

$$TPNH_2 + DPN \rightleftarrows TPN + DPNH_2. \qquad (6, 9)$$

Whilst the presence of this enzyme is certain, its efficiency under physiological conditions is not clear. GLOCK and McLEAN [1955 (a), (b)] found that $TPNH_2$ is prevented in the intact tissues from reacting freely according to (6, 9), perhaps because the reactants are largely bound to other cell constituents, or because they are inaccessible to each other in separate cell compartments.

Experiments with isotopically labelled glucose suggest that in animal tissues or yeast only a small proportion of the total glucose breakdown is accounted for by the pentose phosphate cycle, but owing to difficulties of experimentation, inherent in the cyclic nature of the processes concerned, it has not yet been possible to achieve a precise quantitative evaluation of the contributions made by the different pathways of carbohydrate breakdown (see WOOD 1955).

Apart from the pentose phosphate cycle, there are several other alternative pathways of carbohydrate degradation. None of these is of such universal occurrence as the pentose phosphate cycle; these other pathways seem to be restricted to a limited range of micro-organisms (see GUNSALUS, HORECKER & WOOD 1955, RACKER 1954).

7. The Path of Carbon in Photosynthesis

The pentose phosphate cycle of glucose oxidation has many features in common with the chemical reactions by which green plants (and probably other autotrophic organisms, see SANTER & VISHNIAC 1955, TRUDINGER 1955) convert carbon dioxide to sugar.

In effect, photosynthesis is the reversal of respiration, in so far as oxygen is evolved, carbon dioxide is absorbed and carbohydrate formed. However, the sequence of reactions in photosynthesis is bound to be different from that in respiration, at least in part, because the synthesis of carbohydrate, being endergonic, must be linked with exergonic reactions. The pathway of carbon in photosynthesis is known to include 15 different reactions: this is still a minimum figure, as several stages are probably complex and could be further resolved. Of these 15 reactions only 2 are specific to photosynthetic and autotrophic organisms. All others also occur in non-photosynthetic cells, either as steps of the pentose phosphate cycle, or of glycolysis, of or carbohydrate synthesis from noncarbohydrate precursors. Some of the reactions are reversible, and the direction in photosynthesis is the reverse of that in glycolysis or of that in the pentose phosphate cycle.

The first of the two reactions of carbon compounds specific for photosynthetic and autotrophic organisms is the phosphorylation of ribulose 5-phosphate to form ribulose 1:5-diphosphate [WEISSBACH, SMYRNIOTIS and HORECKER

1954(a), (b), Hurwitz 1955, Hurwitz et al. 1956, Horecker, Hurwitz and Weissbach 1956].

$$\begin{array}{c} CH_2OH \\ | \\ C=O \\ | \\ H-C-OH \\ | \\ H-C-OH \\ | \\ CH_2OPO_3H_2 \end{array} + ATP \longrightarrow \begin{array}{c} CH_2OPO_3H_2 \\ | \\ C=O \\ | \\ H-C-OH \\ | \\ H-C-OH \\ | \\ CH_2OPO_3H_2 \end{array} + ADP \qquad (7, 1)$$

ribulose 5-phosphate   *ribulose 1:5-diphosphate*

This reaction is catalysed by phosphoribulokinase (Hurwitz 1955, Hurwitz et al. 1956), and is analogous to the action of phosphofructokinase in glycolysis, by which fructose 6-phosphate is converted to fructose 1:6-diphosphate.

The second specific reaction is the cleavage of ribulose 1:5-diphosphate by carbon dioxide and water, to yield 2 molecules of 3-phosphoglyceric acid (Quayle, Fuller, Benson and Calvin 1954, Weissbach and Horecker 1955, Weissbach, Horecker and Hurwitz 1956, Jakoby, Brummond and Ochoa 1956), catalysed by "carboxydismutase" (Calvin, Quayle, Fuller, Mayaudon, Benson and Bassham 1955):

$$\begin{array}{c} HO \\ | \\ O=C \\ | \\ HO \end{array} + \begin{array}{c} CH_2OPO_3H_2 \\ | \\ C=O \\ | \\ H-C-OH \\ | \\ H-C-OH \\ | \\ CH_2OPO_3H_2 \end{array} \longrightarrow \begin{array}{c} CH_2OPO_3H_2 \\ | \\ H-C-OH \\ | \\ O{=}C{-}OH \\ + \\ O{=}C{-}OH \\ | \\ H-C-OH \\ | \\ CH_2OPO_3H_2 \end{array} \qquad (7, 2)$$

carbonic acid + ribulose 1:5-diphosphate   *2 moles 3-phosphoglyceric acid*

The enzyme is referred to as "carboxydismutase" because the reaction can be looked upon as a coupled oxidoreduction in which CO_2 is reduced to carboxyl, a ketone is reduced to a secondary alcohol, and another secondary alcohol is oxidized to carboxyl.

The 3-phosphoglycerate formed by reaction (7, 2) undergoes changes which are identical with those of glycolysis, except that they proceed in the reverse direction and that DPN may be replaced by TPN (Arnon, Rosenberg and Whatley 1954). 3-phosphoglycerate is first phosphorylated to 1:3-diphosphoglycerate in a reaction (5, 2), catalysed by phosphoglycerate kinase and previously described,

$$\begin{array}{c} O \\ \diagdown \\ C-OH \\ | \\ H-C-OH \\ | \\ CH_2OPO_3H_2 \end{array} + ATP \rightarrow \begin{array}{c} O \\ \diagdown \\ C-OPO_3H_2 \\ | \\ H-C-OH \\ | \\ CH_2OPO_3H_2 \end{array} + ADP \qquad (5, 2)$$

3-phosphoglyceric acid   *1:3-diphosphoglyceric acid*

and then reacts with DPNH$_2$ or TPNH$_2$ and triose phosphate dehydrogenase [cf. (5, 3)] to form glyceraldehyde 3-phosphate (7, 3).

$$\begin{array}{c} \text{O} \\ \diagdown \\ \text{CO—PO}_3\text{H}_2 \\ | \\ \text{H—C—OH} \\ | \\ \text{CH}_2\text{OPO}_3\text{H}_2 \end{array} + \text{DPNH}_2 \text{ or TPNH}_2 \rightarrow \begin{array}{c} \text{O} \\ \diagdown \\ \text{C—H} \\ | \\ \text{H—C—OH} \\ | \\ \text{CH}_2\text{OPO}_3\text{H}_2 \end{array} + \text{DPN or TPN} + \text{H}_3\text{PO}_4 \quad (7, 3)$$

2 molecules of glyceraldehyde 3-phosphate are converted to 1 molecule of fructose 1:6-diphosphate by reactions (6, 9) and (6, 10), and the phosphate attached to carbon atom 1 is split off by a fructose 1:6-diphosphatase (6, 11).

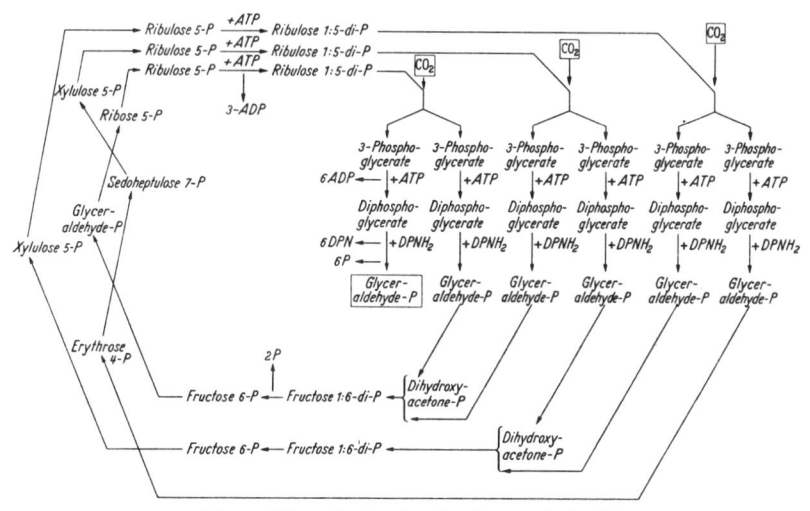

Fig. 9. *The path of carbon in photosynthesis (I)*
(This scheme involves transaldolase, but not sedoheptulose 1:7-diP)

Finally three reactions catalysed by transketolase and transaldolase follow. These are the reactions (6, 7), (6, 6) and (6, 5) of the pentose phosphate cycle, acting in the reverse direction. They lead to the formation of xylulose 5-phosphate. This is then epimerized to ribulose 5-phosphate, by reaction (6, 4). The cycle initiated by (7, 1) is thereby completed.

The quantitative relations of this cycle are conveniently represented by a diagram (Fig. 9). One turn of the cycle, beginning with 3 molecules of ribulose 5-phosphate and 3 molecules of CO$_2$, on balance leads to the formation of 1 molecule of triosephosphate. As formulated, the cycle includes 12 different reactions, and a total of 30 reactions. The latter are listed in Table 8. Two turns of the cycle yield 2 molecules of glyceraldehyde 3-phosphate, which can form a glucose moiety of starch by the six further reactions shown in Fig. 10. This brings the total number of reactions required to form one glucose molecule to 66. The reactions of Fig. 10 are those of glycolysis in reverse, except for the stage

$$\text{fructose 1:6-diP} \rightarrow \text{fructose 6-P}$$

Table 8. *The reactions of the pathway of carbon in photosynthesis*

No.	Equivalents reacting per cycle	Reaction	Enzyme
(1)	3	ribulose 5-P + ATP → ribulose 1:5-diP + ADP	phosphoribulokinase
(2)	3	ribulose 1:5-diP + CO_2 + H_2O → 2 phosphoglyceric acid	carboxydismutase
(3)	6	phosphoglyceric acid + ATP → diphosphoglyceric acid + ADP	phosphoglycerate kinase
(4)	6	diphosphoglyceric acid + $DPNH_2$ → glyceraldehyde 3-P + DPN + P	triosephosphate dehydrogenase
(5)	2	glyceraldehyde 3-P → dihydroxyacetone P	triosephosphate isomerase
(6)	2	glyceraldehyde 3-P + dihydroxyacetone-P → fructose 1:6-diP	aldolase
(7)	2	fructose 1:6-diP + H_2O → fructose 6-P + P	fructose 1:6-diphosphatase
(8)	1	fructose 6-P + glyceraldehyde 3-P → xylulose 5-P + erythrose 4-P	transketolase
(9)	1	fructose 6-P + erythrose 4-P → sedoheptulose 7-P + glyceraldehyde 3-P	transaldolase
(10)	2	sedoheptulose 7-P + glyceraldehyde 3-P → xylulose 5-P + ribose 5-P	transketolase
(11)	2	xylulose 5-P → ribulose 5-P	xylulo epimerase
(12)	1	ribose 5-P → ribulose 5-P	pentose phosphate isomerase
Sum:		3 CO_2 + 9 ATP + 5 H_2O + 6 $DPNH_2$ → glyceraldehyde 3-P + 9 ADP + 6 DPN + 8 P	

which is catalysed by the phosphatase already mentioned as being present in photosynthesizing cells [RACKER 1954(c)].

The evidence supporting this concept consists of the demonstration of the occurrence of every stage of the cycle and of the actual accumulation of sugar which occurs when a water-soluble preparation from spinach, supplemented by co-factors and some enzymes of glycolysis, is incubated with CO_2, catalytic amounts of ribulose diphosphate, an excess of ATP, and catalytic amounts of DPN, which is reduced continuously to $DPNH_2$ either with alcohol plus alcohol dehydrogenase or with molecular hydrogen and a DPN-linked hydrogenase (RACKER 1955).

This scheme nevertheless must be looked upon as provisional. There is one observation which it does not explain, namely the presence of sedoheptulose 1:7-diphosphate in photosynthetic plant material and the appearance of radioactivity in this substance when the plant is exposed to radioactive CO_2. This could be accounted for by a minor modification of the scheme, i.e. by the assumption that reactions 6 and 7 of Table 8 occur only

```
                glyceraldehyde 3-P      glyceraldehyde 3-P
                       |                        | (triosephosphate isomerase)
                       |                        ↓
                       |                dihydroxyacetone P
                       |                        | (aldolase)
                       |_____|
                                    ↓
                            fructose 1:6-diP
                                    | (fructose 1:6-diphosphatase)
                            fructose 6-P + P
                                    | (hexosephosphate isomerase)
                            glucose 6-P
                                    | (phosphoglucomutase)
                            glucose 1-P
                                    | (phosphorylase)
                            starch
```

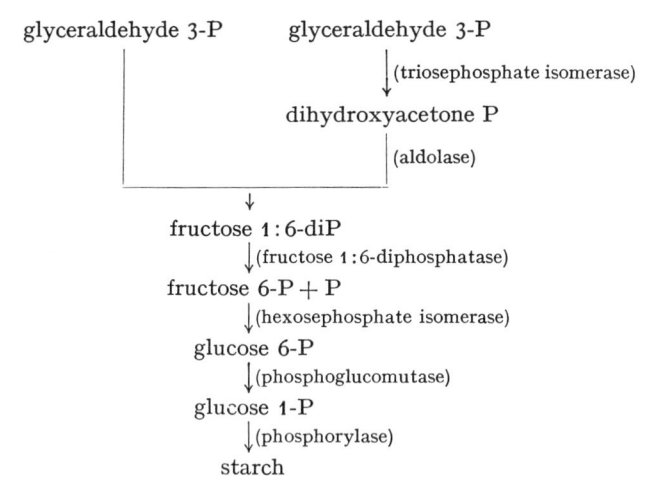

Fig. 10. *Pathway of synthesis of starch from glyceraldehyde*

once instead of twice and that reaction 9 is replaced by the two new reactions (7, 4) and (7, 5).

$$\begin{array}{c} H-C=O \\ | \\ H-C-OH \\ | \\ H-C-OH \\ | \\ CH_2OPO_3H_2 \end{array} \;+\; \begin{array}{c} CH_2OH \\ | \\ C=O \\ | \\ CH_2OPO_3H_2 \end{array} \longrightarrow \begin{array}{c} CH_2OPO_3H_2 \\ | \\ C=O \\ | \\ HO-C-H \\ | \\ H-C-OH \\ | \\ H-C-OH \\ | \\ H-C-OH \\ | \\ CH_2OPO_3H_2 \end{array} \quad (7,4)$$

erythrose 4-phosphate   *dihydroxyacetone phosphate*   *sedoheptulose 1:7-diphosphate*

In reaction (7, 4), catalysed by aldolase, sedoheptulose 1:7-diphosphate is formed (HORECKER, SMYRNIOTIS, HIATT and MARKS 1954). This reaction is similar to the formation of fructose 1:6-diphosphate catalysed by aldolase, except that in the present case dihydroxyacetone phosphate reacts with erythrose 4-phosphate instead of with glyceraldehyde 3-phosphate. The sedoheptulose 1:7-diphosphate produced is next (7, 5) postulated to yield sedoheptulose 7-phosphate, by hydrolysis of the phosphate group attached to carbon atom 1

$$\begin{array}{c} CH_2OPO_3H_2 \\ | \\ C=O \\ | \\ HO-C-H \\ | \\ H-C-OH \\ | \\ H-C-OH \\ | \\ H-C-OH \\ | \\ CH_2OPO_3H_2 \end{array} \longrightarrow \begin{array}{c} CH_2OH \\ | \\ C=O \\ | \\ HO-C-H \\ | \\ H-C-OH \\ | \\ H-C-OH \\ | \\ H-C-OH \\ | \\ CH_2OPO_3H_2 \end{array} \;+\; H_3PO_4 \quad (7,5)$$

sedoheptulose 1:7-diphosphate   *sedoheptulose 7-phosphate* under the action of a phosphatase. Again, this reaction is similar to the conversion of fructose 1:6-diphosphate to fructose 6-phosphate already described.

The net effect of this modification is the same as that of reactions 6, 7 and 9 of Table 8, and the balance of all reactions remains therefore as stated at the bottom of that table. The modified scheme as a whole is shown in Fig. 11. It will be noted that the action of transaldolase is replaced in this modification by that of aldolase. The scheme has the disadvantage that it does not account for the presence of transaldolase in photosynthetic cells. Since both transaldolase and sedoheptulose 1:7-diphosphate have been demonstrated to occur in

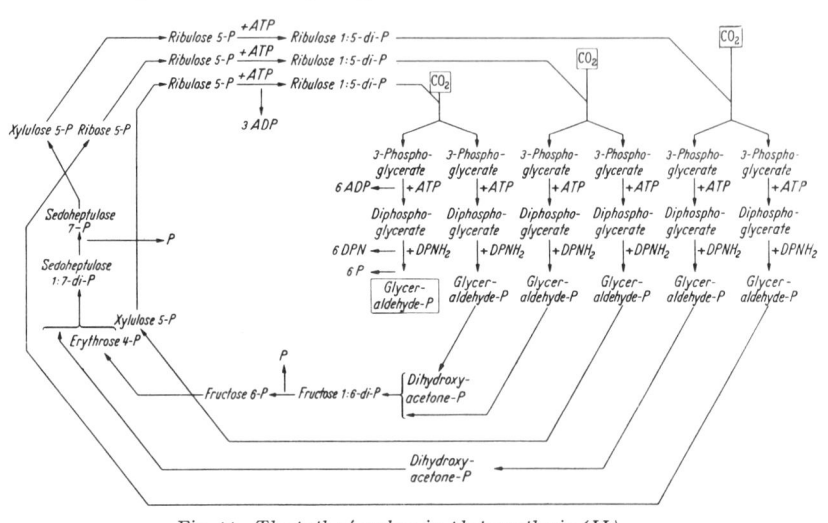

Fig. 11. *The path of carbon in photosynthesis (II)*
(This scheme involves sedoheptulose 1:7-diphosphate, but not transaldolase)

photosynthetic cells, and since the fructose 1:6-diphosphatase present in such cells also catalyses reaction (7, 5) [RACKER 1954(c)], it may be that both pathways represented in Figs. 9 and 11 occur concurrently.

The cycle requires ATP and reduced DPN or TPN in order to proceed. These two reactants arise in another component of the photosynthetic mechanism. The energy needed for the synthesis of these substances from ADP, inorganic phosphate and DPN is in the last resort derived from the absorbed light. The nature of these reactions cannot be discussed in this survey (see CALVIN 1955, ARNON 1955).

A remarkable feature of this pathway is the fact that there is only one reaction in which carbon dioxide is fixed, and only one reaction where a reduction takes place. The latter is especially noteworthy because the reduction of CO_2 to $\overset{|}{C}H(OH)$ is equivalent to the addition of 4 protons. The reduction of phosphoglyceric acid to glyceraldehyde 3-phosphate occurs in fact twice for each molecule of CO_2 fixed. In addition to this step other reactions which amount to coupled oxidoreductions occur. One of these is the "carboxydismutase" reaction (7, 2) already discussed. Another is the aldol condensation of two triose phosphates when an aldehyde is reduced to a secondary alcohol whilst a primary alcohol is oxidized to a secondary one.

It is true that a second carboxylation reaction takes place in photosynthetic cells, namely the formation of malic acid from phosphoglycerate:

$$\text{phosphoglycerate} \rightarrow \text{phosphopyruvate} \xrightarrow{+CO_2} \text{oxaloacetate} \rightarrow \text{malate.}$$

This reaction is not restricted to photosynthetic organisms; it is widespread in animal tissues and in bacteria. It is not a stage in the conversion of CO_2 to sugar but contributes towards the synthesis of di- and tri-carboxylic acids and amino acids from phosphoglycerate.

8. Utilization of Energy for Chemical Syntheses

A general treatment of the mechanism by which energy, liberated in the degradation of foodstuffs, is utilized to feed energy-requiring processes is beyond the scope of this survey. It is well established that ATP plays a key role as an energy transmitter in muscular contraction (WEBER 1954, 1955, WEBER and PORTZAHL 1954, MORALES, BOTTS, BLUM and HILL 1955) and probably of protoplasmatic movement generally (WEBER 1954). It is a fuel in the generation of light in the lantern of the firefly [though not in every case of bioluminescence (McELROY 1951, McELROY and STREHLER 1954)] and of electric currents in the electric organ and in nerve tissue (NACHMANSOHN 1955, NACHMANSOHN et al. 1943, 1946). It plays a role in "active" transport of solutes in secretion and absorption (see DAVIES 1954). These aspects of the fuel function of ATP have been reviewed elsewhere.

It is proposed here to discuss the mechanisms by which chemical syntheses are achieved at the expense of the free energy of degradative reactions, an aspect of energy utilization which has not been surveyed elsewhere.

Synthesis is defined in the present context as any chemical reaction in which free energy is lost. It is evident that syntheses in this sense are always "half-reactions". They can only take place if coupled with another chemical reaction in which energy is liberated. The nature of the coupling mechanisms by which the endergonic and exergonic reactions are so linked as to make a transfer of energy possible is in fact the main problem to be considered. The coupling mechanisms employed in the biosyntheses of are various types. In most cases ATP takes part at some stages, but there are also syntheses which do not involve ATP. Four cases illustrating the different types of mechanisms are selected for detailed discussion.

Case I. Synthesis of glycogen from glucose

A case where the coupling mechanism is simple is that of glycogen synthesis from glucose:

$$\text{glucose} \rightarrow \text{glycogen} \ (\Delta G = +7.0 \text{ kgcal})^1 \qquad (8, 1)$$

This endergonic reaction is coupled with the exergonic reaction

$$\text{ATP}^{4-} + \text{H}_2\text{O} \rightarrow \text{ADP}^{3-} + \text{P}^{2-} + \text{H}^+ \ (\Delta G = -11.6 \text{ kgcal}) \qquad (8, 2)$$

in the following manner:

$$\text{glucose} + \text{ATP} \rightarrow \text{glucose 6-phosphate} + \text{ADP}$$
$$\text{glucose 6-phosphate} \rightarrow \text{glucose 1-phosphate}$$
$$\text{glucose 1-phosphate} \rightarrow \text{glycogen} + \text{P}.$$

The sum of these three reactions is (8, 1) and (8, 2). Coupling is thus achieved by the introduction of intermediary steps in which the reactants of both the endergonic and exergonic reactions take part. Each individual step is exergonic (at suitable concentrations of the reactants), as is the sum of (8, 1) and (8, 2), and it is clear how any metabolic reaction leading to the synthesis of ATP can energize the synthesis of polysaccharide from glucose.

Case II. Synthesis of carbohydrate from lactate

A more complex case is the synthesis in liver tissue of carbohydrate from non-carbohydrate precursors such as lactate or pyruvate. Whilst this synthesis is in effect the reversal of glycolysis, the enzymic mechanism cannot, for thermodynamic reasons, be a simple reversal of glycolysis. The changes associated with the fermentation of a glucose equivalent of glycogen may be formulated as follows:

$$\text{glycogen} \rightarrow 2 \text{ lactate}^- + 2\text{H}^+ \ (\Delta G = -57.0 \text{ kgcal}), \qquad (8, 3)$$

$$3 \ [\text{ADP}^{3-} + \text{HPO}_4^{2-} + \text{H}^+ \rightarrow \text{ATP}^{4-} + \text{H}_2\text{O}] \ (\Delta G = +3 \times 11.6 \text{ kgcal}). \qquad (8, 4)$$

The sum of these two reactions is

$$\text{glycogen} + 3 \text{ ADP}^{3-} + 3 \text{ HPO}_4^{2-} + 3\text{H}^+ \rightarrow 2 \text{ lactate}^- + 2 \text{ ATP}^{4-} + 3 \text{ H}_2\text{O} \\ (\Delta G = -22.2 \text{ kgcal}). \qquad (8, 5)$$

If the analogous calculations are made for the fermentation of glucose the following data are obtained:

$$\text{glucose} + 2 \text{ ADP}^{3-} + 2 \text{ HPO}_4^{2-} \rightarrow 2 \text{ lactate}^- + 2 \text{ ATP}^{4-} + \text{H}_2\text{O} \ (\Delta G = -26.9 \text{ kgcal}). \qquad (8, 6)$$

As all steps of glycolysis are reversible, it has been assumed by some authors that carbohydrate can be synthesised from lactate by the simple reversal of (8, 5) and (8, 6). A high concentration of ATP and a low concentration of ADP, achieved by oxidative phosphorylation, was taken to be the required driving force. All but three steps of glycolysis are, in fact, readily reversed, but at

---

[1] The free energy values in this chapter refer to the conditions stated in the heading of Table 2.

these three stages occur major energy barriers which are liable to prevent an appreciable net reversal. Adenosine phosphates are reactants in all three:

$$\text{glucose} + \text{ATP}^{4-} \rightarrow \text{glucose 6-P}^{2-} + \text{ADP}^{3-} + \text{H}^+, \quad (8, 7)$$

$$\text{fructose 6-P}^{2-} + \text{ATP}^{4-} \rightarrow \text{fructose 1:6-diP}^{4-} + \text{ADP}^{3-} + \text{H}^+, \quad (8, 8)$$

$$\text{phosphopyruvate}^{3-} + \text{ADP}^{3-} + \text{H}^+ \rightarrow \text{pyruvate}^- + \text{ATP}^{4-}. \quad (8, 9)$$

Reaction (8, 9), it should be borne in mind, must occur twice for each molecule of carbohydrate and most of the total energy barrier of 26.9 kgcal required to synthesise one molecule of glucose from 2 molecules of lactate is thus to be overcome mainly in four individual reactions, each of which is endergonic (when proceeding from left to right) by about 5—6 kgcal. Concentrations of reactants which would favour such reversals can be calculated from the equation

$$\Delta G = \Delta G^0 + RT \ln \frac{\text{Product of concentrations of end products}}{\text{Product of concentrations of starting materials}}.$$

The calculations show that a high ratio of ATP/ADP is needed for the reversal of (8, 9) (about 10^6 if the other reactants are present at equal concentrations), whilst a low ratio is needed for the reversal of (8, 7) and (8, 8) (about 10^{-6} if the other reactants are present at equal concentrations). The two conditions can hardly exist side by side in cells performing the synthesis, even if the two reactions occur at different sites. A simple reversal of fermentation must therefore be discounted as a mechanism of *net* synthesis. The evidence indicates that the pathway of carbohydrate synthesis from lactate uses those steps of glycolysis which are readily reversible but includes special reactions which circumvent the energy barriers at the stages (8, 7), (8, 8) and (8, 9). In the cases of reactions (8, 7) and (8, 8), the barrier can be circumvented if, in the synthetic pathways, the phosphate of the hexose phosphate esters is removed by hydrolysis rather than by transfer to ATP; i.e. if the reversal of reaction (8, 7) is replaced by

$$\text{fructose 1:6-diphosphate} + \text{H}_2\text{O} \rightarrow \text{fructose 6-phosphate} + \text{phosphate} \quad (8, 10)$$

and the reversal of reaction (8, 8) by

$$\text{glucose 6-phosphate} + \text{H}_2\text{O} \rightarrow \text{glucose} + \text{phosphate}. \quad (8, 11)$$

The two special enzymes required for (8, 10) and (8, 11) do in fact occur at the sites of glycogen synthesis. These are the specific fructose 1:6-diphosphatase, releasing one phosphate group of fructose 1:6-diphosphate, which has been already mentioned [see reaction (6, 11) (GOMORI 1943, POGELL and MCGILVERY 1954)], and a specific glucose 6-phosphatase (FANTL & ROME 1945, SWANSON 1950, CORI & CORI 1952).

The energy barrier presented by the reversal of reaction (8, 9) is circumvented by a more complex mechanism, which includes the three reactions

(8, 12) to (8, 14):

$$\text{pyruvate} + \text{TPNH}_2 + \text{CO}_2 \rightarrow \text{malate} + \text{TPN} \qquad (8, 12)$$
(catalysed by the "malic enzyme" of OCHOA, MEHLER & KORNBERG 1948),

$$\text{malate} + \text{DPN} \rightarrow \text{oxaloacetate} + \text{DPNH}_2 \qquad (8, 13)$$
(catalysed by malic dehydrogenase),

$$\text{oxaloacetate} + \text{ITP} \rightarrow \text{phosphopyruvate} + \text{CO}_2 + \text{IDP} \qquad (8, 14)$$
[catalysed by the enzyme of UTTER and KURAHASHI 1954).

The sum of reactions (8, 12) to (8, 14) is

$$\text{pyruvate} + \text{TPNH}_2 + \text{DPN} + \text{ITP} \rightarrow \text{phosphopyruvate} + \text{DPNH}_2 + \text{TPN} + \text{IDP}. \quad (8, 15)$$

Since the oxidation-reduction potential of the DPNH_2/DPN and TPNH_2/TPN systems are virtually identical, and since there are no major differences between the free energies of hydrolysis of ITP and ATP, reaction (8, 15) must be no less endergonic than is reaction (8, 9) when going from right to left, i.e. ΔG must be $+ 6.0$ kgcal. But, whilst the single reaction (8, 9) presents an energy barrier, the similar energy barrier presented by the over-all reaction (8, 15) can be surmounted by coupling the component reactions (8, 12) to (8, 14), with exergonic reactions. Any such coupled reaction which produces TPNH_2, DPN or ITP will facilitate the production of phosphopyruvate. Thus, the formation of malate from pyruvate [reaction (8, 12)] can be coupled with one of the TPNH_2-producing dehydrogenase systems, such as the glucose 6-phosphate dehydrogenase (6, 1) or 6-phosphogluconate dehydrogenase (6, 2) systems previously described, or the isocitrate dehydrogenase reaction

$$\text{isocitrate} + \text{TPN} \rightarrow \alpha\text{-ketoglutarate} + \text{CO}_2 + \text{TPNH}_2. \qquad (8, 16)$$

Of these, reaction (8, 16), although it has been demonstrated *in vitro*, cannot play a major part in normal carbohydrate synthesis. This is, because for each molecule of malate formed by (8, 12), one molecule of isocitrate would have to be oxidized. Since each molecule of isocitrate arising in the tricarboxylic acid cycle represents a stage in the oxidation of half an equivalent of glucose, it would require one molecule of glucose to be oxidized for one molecule of glucose to be synthesized, were (8, 12) coupled *in vivo* only with (8, 16). The coupling of reaction (8, 12) with the oxidative reactions of the pentose phosphate cycle, (6, 1) and (6, 2), is quantitatively feasible, if it is assumed that the pentose phosphate cycle operates rapidly enough to regenerate the requisite amounts of glucose 6-phophate. The available evidence suggests, however, that yet another mode of coupling of (8, 12) with TPNH_2-producing reactions operates. If the pentose phosphate cycle provided TPNH_2 at a sufficiently rapid rate, maximum rates of malate synthesis should occur anaerobically on addition of glucose 6-phosphate. Such is not the case: in pigeon liver preparations, oxygen is far more effective in promoting reaction (8, 12) than is glucose 6-phosphate under anaerobic conditions [KREBS 1954(c)]. Since only small amounts of lactate are formed in this system, the oxygen effect cannot be due to the re-oxidation of lactate. Moreover, the stimulation by oxygen is abolished by addition of 2:4-dinitrophenol. These findings suggest a production of TPNH$_2$ by a reversal of oxidative phosphorylation, of the type

reduced flavoprotein + TPN + ATP $\rightleftarrows$ flavoprotein + TPNH$_2$ + ADP + P

discussed more fully in Chapter 10.

Reaction (8, 13) can be coupled with the oxidation of reduced pyridine nucleotide by the flavoprotein-cytochrome-O$_2$ system. Not only is DPNH$_2$ reoxidized to DPN in this reaction (3, 5) [thus facilitating reaction (8, 13)], but also, by coupled oxidative phosphorylation, ATP is produced from ADP. Since ATP and ITP are interconvertible by reaction (8, 17)

$$\text{ATP} + \text{IDP} \rightleftarrows \text{ADP} + \text{ITP} \tag{8, 17}$$
(KREBS and HEMS 1953, BERG and JOKLIK 1953, 1954)

the oxidation of reduced pyridine nucleotide facilitates both reactions (8, 13) and (8, 14).

Whilst, then, the overall reaction (8, 15) is endergonic, the coupling of its constituent reactions (8, 12) to (8, 14) with further reactions renders the system exergonic.

These considerations also make it possible to suggest a stoichiometric explanation for the maximum values of the original "MEYERHOF quotient"

$$\frac{\text{molecules of lactate resynthesised to carbohydrate}}{\text{molecules O}_2 \text{ used for oxidation of lactate}}.$$

(This should not be confused with another related metabolic quotient, also referred to in the literature as "MEYERHOF quotient":

$$\frac{\text{anaerobic lactic fermentation} - \text{aerobic lactic fermentation}}{\text{oxygen uptake}} \Bigg).$$

The synthesis of carbohydrate from lactate involves two dehydrogenations [in the conversion of lactate to pyruvate, and in the reaction coupled with reaction (8, 12)] and two hydrogenations (in the conversion of pyruvate to malate and of phosphoglycerate to triose phosphate). The suggested coupling of (8, 13) with the oxidation of DPNH$_2$ to DPN by oxygen would thus leave a deficit of one molecule of DPNH$_2$ in the overall scheme. This is met if another substrate molecule, for example lactate, is oxidized by DPN:

$$6 \text{ DPN} + \text{lactate} \rightarrow 6 \text{ DPNH}_2 + 3 \text{ CO}_2. \tag{8, 18}$$

In this case one molecule of lactate would be oxidized for six molecules of lactate converted to carbohydrate. This was found by MEYERHOF (1920) to be the case.

Fig. 12 shows the pathway of carbohydrate synthesis from lactate as a whole, without the coupled reactions at the stages between pyruvate and phosphopyruvate. The arrows on the right represent the synthesis, those on the left glycolysis. Every intermediate carbon compound of glycolysis occurs also in the synthesis, but the reactions by which the intermediates are formed differ at three points. Another way of describing the situation is to say that cyclic systems, in which the interconversion of the products involve different mechanisms for the backward and forward reactions (see KREBS 1947), are inserted at three points in the glycolytic chain of reactions.

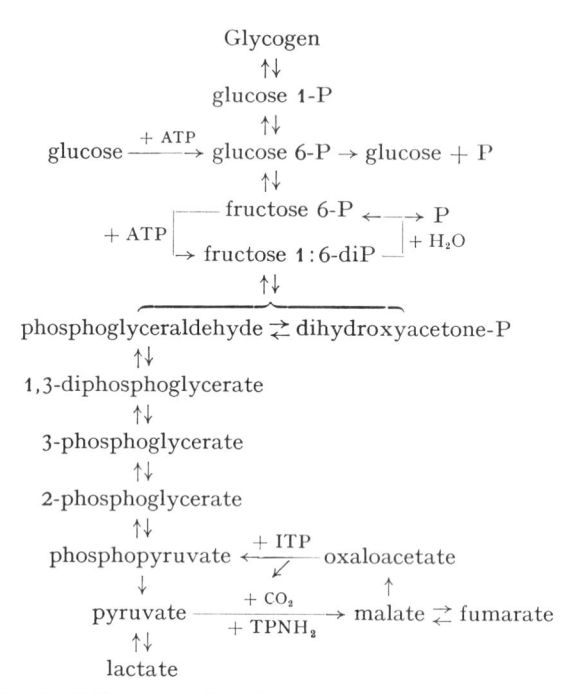

$$
\begin{array}{c}
\text{Glycogen} \\
\updownarrow \\
\text{glucose 1-P} \\
\updownarrow \\
\text{glucose} \xrightarrow{+\text{ATP}} \text{glucose 6-P} \rightarrow \text{glucose} + \text{P} \\
\updownarrow \\
+\text{ATP} \begin{bmatrix} \text{fructose 6-P} & \xleftarrow{}\rightarrow \text{P} \\ \rightarrow \text{fructose 1:6-diP} \end{bmatrix} +\text{H}_2\text{O} \\
\updownarrow \\
\overbrace{\text{phosphoglyceraldehyde} \rightleftarrows \text{dihydroxyacetone-P}} \\
\updownarrow \\
\text{1,3-diphosphoglycerate} \\
\updownarrow \\
\text{3-phosphoglycerate} \\
\updownarrow \\
\text{2-phosphoglycerate} \\
\updownarrow \\
\text{phosphopyruvate} \xleftarrow{+\text{ITP}} \text{oxaloacetate} \\
\downarrow \uparrow \\
\text{pyruvate} \xrightarrow[+\text{TPNH}_2]{+\text{CO}_2} \text{malate} \rightleftarrows \text{fumarate} \\
\updownarrow \\
\text{lactate}
\end{array}
$$

Fig. 12. *Pathways of carbohydrate breakdown and synthesis*
(The pathways differ at three points. Catabolic reactions are indicated by the left hand arrows, anabolic reactions by the right hand arrows)

All the postulated reactions have been demonstrated to occur. This establishes the possibility of the mechanism. That carbohydrate is actually formed through the postulated sequence is borne out by two independent facts. The first is the participation of carbon dioxide in the synthesis of sugar. When glycogen is synthesized in the liver in the presence of labelled CO_2, the label appears in carbons 3 and 4 of the glucose moieties. That this is not a minor exchange reaction is shown by the quantitative measurements of TOPPER & HASTINGS (1949). Their calculations indicate that a minimum of 84% of the pyruvate molecules must have fixed CO_2, by a reaction of the type (8, 12), before they became part of the sugar molecule in the liver. Secondly, if pyruvate entered the glycolytic chain by reversal of the pyruvate kinase reaction (8, 9), isotope from pyruvate or lactate labelled with ^{14}C in the α-carbon should appear only in carbons 2 and 5 of the glucose moieties of glycogen, whilst β-labelled pyruvate and lactate should yield glucose with the isotope only in carbons 1 and 6. In fact, however, the label is uniformly distributed over carbons 1, 2, 5 and 6 of glucose, irrespective of whether α or β labelled pyruvate or lactate is administered to the animal (TOPPER & HASTINGS 1949, LORBER, LIFSON, WOOD, SAKAMI and SHREEVE 1950, LORBER, LIFSON, SAKAMI and WOOD 1950). This means that the α and β carbon atoms must at one stage be part of a symmetrical molecule. The present concept makes provision for this, the symmetrical molecule being fumarate, formed by a rapid side reaction from malate.

Case III. Synthesis of Fatty Acids

Long chain fatty acids are known to be broken down and built up by the removal or addition of C_2-units reacting in the form of acetyl coenzyme A. There are four reversible reactions for each molecule of acetyl coenzyme A, in both breakdown and synthesis. In the case of the breakdown the first step is the dehydrogenation of the fatty acid chain in the α-β position leading to the formation of an α-β double bond, the immediate hydrogen acceptor being a flavoprotein:

$$\begin{array}{c} R \\ | \\ (CH_2)_n \\ | \\ CH_2 \\ | \\ CH_2 \\ | \\ CH_2 \\ | \\ CO \\ | \\ S\text{—}CoA \end{array} + \text{flavoprotein} \rightleftarrows \begin{array}{c} R \\ | \\ (CH_2)_n \\ | \\ CH_2 \\ | \\ CH \\ \| \\ CH \\ | \\ CO \\ | \\ S\text{—}CoA \end{array} + \text{dihydroflavoprotein} \qquad (8, 19)$$

acyl coenzyme A   α-β-unsaturated acyl coenzyme A

(α-β dehydrogenation of fatty acids and hydrogenation of unsaturated fatty acids, catalysed by "acyl-dehydrogenase" [GREEN 1954(b), 1955] or "ethylene reductase" (LYNEN 1954). According to LANGDON (1955) the reaction from right to left may require $TPNH_2$.)

The second step is the hydration of the double bond leading to a β-hydroxy acid.

$$\begin{array}{c} R \\ | \\ (CH_2)_n \\ | \\ CH_2 \\ | \\ CH \\ \| \\ CH \\ | \\ CO \\ | \\ S\text{—}CoA \end{array} \underset{- H_2O}{\overset{+ H_2O}{\rightleftarrows}} \begin{array}{c} R \\ | \\ (CH_2)_n \\ | \\ CH_2 \\ | \\ CHOH \\ | \\ CH_2 \\ | \\ CO \\ | \\ S\text{—}CoA \end{array} \qquad (8, 20)$$

α-β unsaturated acyl coenzyme A   β-hydroxy acyl coenzyme A

[Catalysed by "crotonase" (SEUBERT & LYNEN 1953, STERN & DEL CAMPILLO 1953) or "enol hydrase" (GREEN 1955).]

This is followed by the dehydrogenation of the β-hydroxy acid to a β-ketonic acid, with DPN as the primary hydrogen acceptor:

$$\begin{matrix} R \\ | \\ (CH_2)_n \\ | \\ CH_2 \\ | \\ CHOH \\ | \\ CH_2 \\ | \\ CO \\ | \\ S-CoA \end{matrix} + DPN \rightleftarrows \begin{matrix} R \\ | \\ (CH_2)_n \\ | \\ CH_2 \\ | \\ CO \\ | \\ CH_2 \\ | \\ CO \\ | \\ S-CoA \end{matrix} + DPNH_2 \quad (8, 21)$$

β-hydroxy acyl coenzyme A   β-keto acyl coenzyme A
(Catalysed by β-hydroxy acyl dehydrogenase.)

The fourth step is the thiolysis of the β-ketonic acid by a molecule of coenzyme A, to yield acetyl coenzyme A and an acyl coenzyme A with two carbon atoms less than entered reaction (8, 19):

$$\begin{matrix} R \\ | \\ (CH_2)_n \\ | \\ CH_2 \\ | \\ CO \\ | \\ CH_2 \\ | \\ CO \\ | \\ S-CoA \end{matrix} + \begin{matrix} CoA \\ | \\ S \\ | \\ H \end{matrix} \rightleftarrows \begin{matrix} R \\ | \\ (CH_2)_n \\ | \\ CH_2 \\ | \\ CO \\ | \\ S-CoA \end{matrix} + \begin{matrix} CH_3 \\ | \\ CO \\ | \\ S-CoA \end{matrix} \quad (8, 22)$$

β-keto acyl coenzyme A   acyl coenzyme A + acetyl coenzyme A
(Catalysed by β-ketothiolase.)

In reverse these four reactions lead to the elongation of fatty acids. As the fatty acids react in all cases as coenzyme A derivatives, one additional reaction is necessary for both degradation and synthesis, namely a step initiating the degradation, and a step completing the synthesis. The reaction which initiates the degradation involves ATP and converts free fatty acid into the corresponding acyl coenzyme A compound according to the general equation

fatty acid + coenzyme A + ATP ⇌ acyl coenzyme A + AMP + pyrophosphate. (8, 23)

This reaction, already referred to as the sum of reactions (4, 9) and (4, 8) proceeding from right to left, has been established for acetic and propionic acids (CHOU & LIPMANN 1952, BEINERT et al. 1953), as well as for medium (MAHLER, WAKIL and BOCK 1953) and long chain fatty acids [KORNBERG & PRICER

1953 (a)]. Although the reaction is reversible it is probable that only the reactions leading to the synthesis of acyl coenzyme A are of major physiological importance (see p. 225). The final stage in the synthesis is the esterification of fatty acids. Information on this stage is still incomplete. According to KORNBERG & PRICER [1953(b)], long chain acyl coenzyme A can react with α-glycerophosphate.

$$\begin{array}{c} CH_2OH \\ | \\ CHOH \\ | \\ CH_2OPO_3H_2 \end{array} + 2\,Acyl\,CoA \rightarrow \begin{array}{c} CH_2O\text{—}Acyl \\ | \\ CHO\text{—}Acyl \\ | \\ CH_2OPO_3H_2 \end{array} \qquad (8, 24)$$

α-glycerophosphate   phosphatidic acid

How the phosphatidic acid is converted into neutral fat, or whether there is a direct pathway leading from glycerol and acyl coenzyme A to neutral fat, is not yet known (see TIETZ and SHAPIRO 1956).

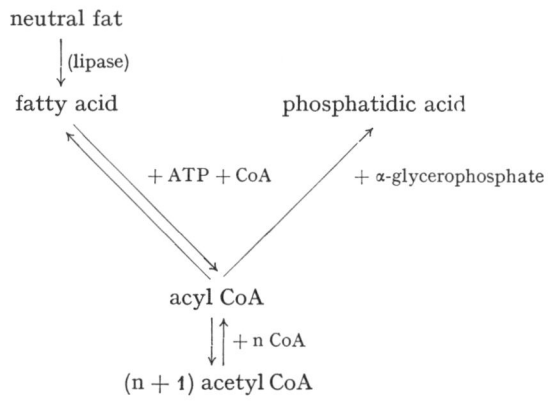

Fig. 13. *The breakdown and synthesis of fat*

The reactions of long chain acyl coenzyme A, as far as they have been discussed, are summarised in Fig. 13. Whilst the lengthening and shortening of acyl coenzyme A chains is reversible, the formation of acyl coenzyme A from fat, and the conversion of acyl coenzyme A to neutral fat, are assumed to require different routes.

The question arises of what determines whether synthesis or degradation occurs in the reversible system. The sum of the four reactions causing shortening or lengthening of fatty acid chains is as follows:

$$C_n\text{-acyl CoA} + DPNH_2 + \text{reduced flavoprotein} + \text{acetyl CoA} \rightleftarrows \\ C_{n+2}\text{-acyl CoA} + DPN + \text{flavoprotein} + CoA \qquad (8, 25)$$

If this reactions is reversible, it proceeds from left to right when the values for the three ratios

$$\frac{\text{acetyl CoA}}{\text{CoA—SH}}, \qquad \frac{\text{reduced flavoprotein}}{\text{oxidized flavoprotein}}, \qquad \frac{DPNH_2}{DPN}$$

are relatively high, and from right to left when the ratios are relatively low. In the intact body, fat is generally synthesised when there is surplus of carbohydrate, i.e. when pyruvate is available to form more acetyl coenzyme A than is required for energy production by the tricarboxylic acid cycle. Concomitant with the formation of acetyl CoA from pyruvate, occurs the reduction of DPN, and, indirectly, of flavoprotein. Thus surplus of pyruvate raises the three above ratios, whereas lack of carbohydrate lowers them. It can therefore be understood, in general terms, why fatty acids are broken down in the absence and synthesised in the presence of carbohydrate.

Case IV. Synthesis of Phenylalanine

In contrast to the metabolism of fatty acid chains, where syntheses occur by the same reactions as do degradations, is the metabolism of a number of amino acids where synthesis and degradation follow totally different routes. An example of this is the case of phenylalanine. Many micro-organisms, unlike higher animals, can synthesise this amino acid from a variety of carbon compounds and ammonia. The chief stages of the synthesis have recently been established for E. coli, Neurospora crassa and Aerobacter aerogenes, thanks to the work of TATUM et al. (1954), DAVIS (1955), SPRINSON (1955), GILVARG (1955), EHRENSVÄRD (1955) and their collaborators. Most of the information has been obtained by the use of isotopic carbon compounds and especially from the study of the metabolism of mutants deficient in enzymes required for the synthesis. Such a deficiency can lead to the accumulation of intermediates which can be isolated, identified, and examined for metabolic properties. Substances which have been established as intermediates are erythrose 4-phosphate, phosphopyruvate, 5-dehydroshikimic acid, shikimic acid, prephenic acid and phenyl pyruvic acid. The pathway of synthesis is shown in Fig. 14. Though incomplete and in part hypothetical, this scheme may be taken to represent correctly the main stages of the synthesis in the organisms studied, and probably also in other organisms, including plants.

Either glucose or sedoheptulose 1:7-diphosphate can serve as starting material for the synthesis of the carbon skeleton of phenylalanine. Both are precursors of erythrose 4-phosphate and phosphopyruvate (see chapter 6). As seen from Fig. 14, the reactions proceeding from erythrose 4-phosphate and phosphopyruvate include a hydrogenation by $TPNH_2$ (for the reduction of dehydroshikimic acid), a decarboxylation, and several condensations and rearrangements, with and without the elimination of water or phosphate. The balance equation, starting from erythrose 4-phosphate, phosphopyruvic acid and glutamate is a follows:

$$\text{erythrose 4-phosphate} + 2 \text{ phosphopyruvate} + \text{glutamate} + TPNH_2 \rightarrow$$
$$\text{phenylalanine} + \alpha\text{-ketoglutarate} + CO_2 + 5 H_2O + TPN + 3 P. \quad (8, 26)$$

Utilization of Energy for Chemical Syntheses

```
      Glucose                          Glucose
        │ + O₂                           │ via pentose cycle
        ↓                                ↓      (Fig. 7)
2 [COOH · C · (OPO₃H₂) : CH₂]    CHO · CH (OH) · CH(OH) · CH₂OPO₃H₂
    phosphopyruvic acid           erythrose 4-phosphate
```

COOH · CO · CH₂ · CH(OH) · CH(OH) · CH(OH) · CH₂OPO₃H₂

2-keto 3-deoxy 7-phospho D-gluco-heptonic acid, hypothetical intermediate; see Srinivasan, Katagiri & Sprinson (1955).

− H₃PO₄
− H₂O

[catalysed by cell-free extracts of E. coli (Srinivasan et. al. 1955). This reaction is analogous to condensation of phosphopyruvate and HCO₃⁻ leading to oxaloacetate (Utter-Kurahashi reaction (8, 14).]

COOH

5-dehydroshikimic acid

+ TPNH₂

[catalysed by 5-dehydroshikimic reductase (Yaniv & Gilvarg 1955)]

COOH

+ TPN   shikimic acid

− 2 H₂O + pyruvate or phosphopyruvate

prephenic acid

− H₂O
− CO₂

CH₂ · CO · COOH

phenyl pyruvic acid

+ glutamate   (catalysed by transaminase)

CH₂ · CH (NH₂) · COOH

phenylalanine

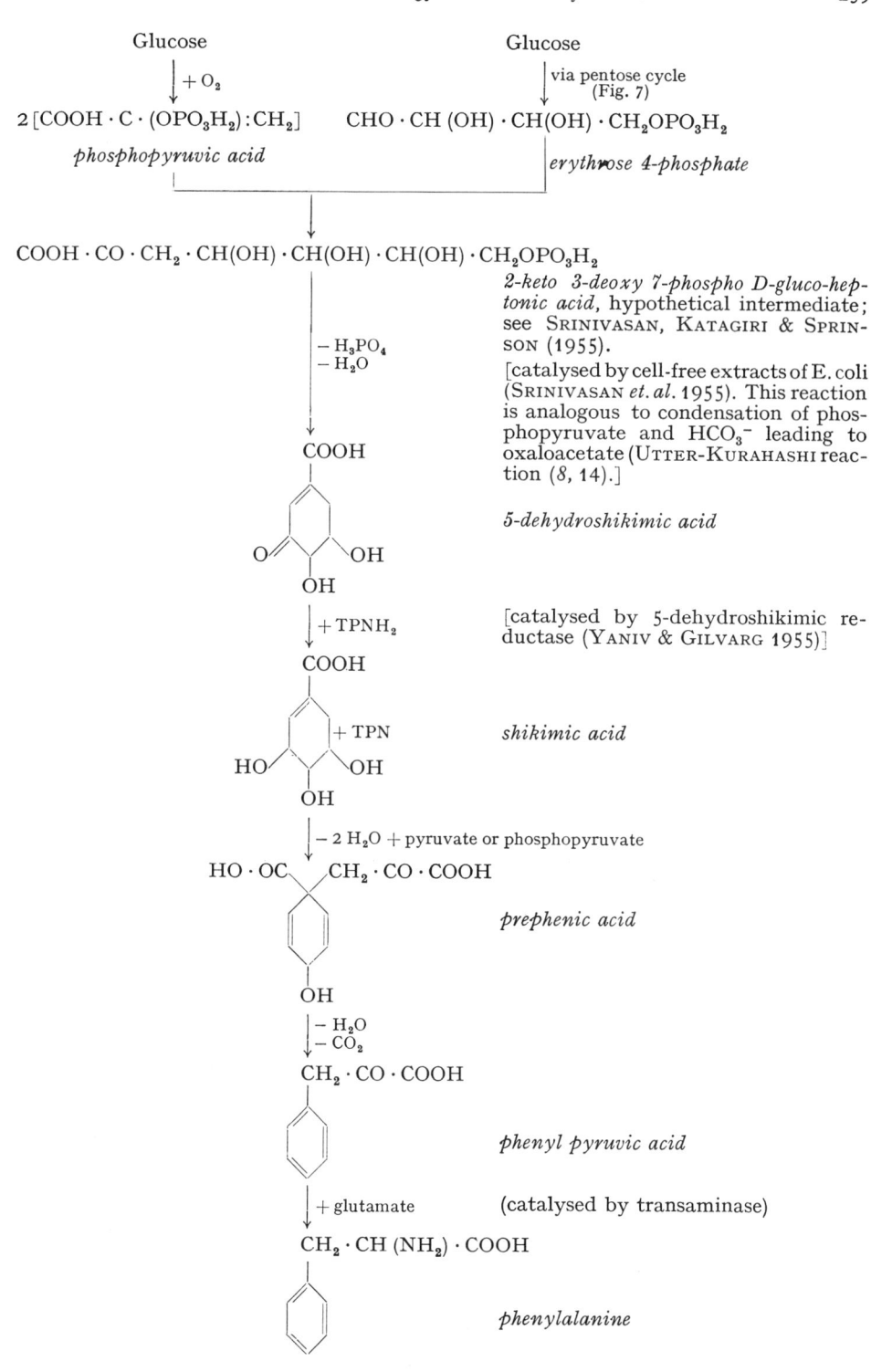

Fig. 14. *Synthesis of phenylalanine*

(The main information is derived from experiments on E. coli and Neurospora. Glucose can be replaced by sedoheptulose 1:7-diphosphate as a source of both erythrose 4-phosphate and phosphopyruvate)

As TPNH$_2$ can be available in only catalytic quantities, the synthesis must be coupled with a TPN-requiring dehydrogenase system. ATP is not necessary for the synthesis when the reactants on the left of reaction (8, 26) are available, because the energy content of the starting material is sufficiently high to allow the over-all synthesis of phenylalanine to be accompanied by a loss of free energy. But ATP is, of course, required for the synthesis of these starting materials. The exergonic nature of most of the intermediary steps of reaction (8, 26) is demonstrated by the fact that they readily occur in solutions containing the required enzymes.

The biological breakdown of phenylalanine passes through entirely different stages, shown in Fig. 15. In the animal, as well as in some micro-organisms, these stages have been shown [see ref. (d) Table 3] to include tyrosine, homogentisic acid, fumaryl-acetoacetic acid, fumaric acid, malic acid and acetoacetic acid.

Exactly analogous situations apply to other amino acids, such as valine, leucine and histidine where the pathways of synthesis and degradation are also entirely different (EHRENSVÄRD 1955).

General considerations

A review of the four examples of synthesis from a general standpoint reveals that ATP participates in all cases, but that the fission of ATP is in most cases not the only mechanism of energy supply. A second source of energy supply is provided by the occurrence of coupled oxido-reductions.

In Case I — synthesis of glycogen from glucose — ATP is the only source of energy. In Case II — synthesis of carbohydrate from lactate — ATP operates in conjunction with coupled oxido-reductions. In Case III — synthesis of fatty acids — ATP does not take part in the majority of steps, (the four reactions concerned with the lengthening of the fatty acid chains) but is required, if free acetate is the precursor of acetyl coenzyme A, through reactions (4, 8) and (4, 9) proceeding from right to left. Since, however, acetyl coenzyme A arises mainly from pyruvate, the synthesis of fatty acids depends on the state of oxido-reduction systems and on the availability of coenzyme A, rather than on the supply of ATP.

The analysis of Case IV shows that ATP, although it does not occur in the reactions of the scheme as presented, plays a necessary role in reactions prior to those listed. The formation of carbohydrate, be it by photosynthesis or by synthesis from lactate, requires ATP. Furthermore, carbohydrate must be converted into glucose 6-phosphate by a reaction requiring ATP before it can yield the precursors of phenylalanine synthesis, erythrose 4-phosphate and phosphopyruvate.

Coupled oxido-reductions may be looked upon as systems where the endergonic and exergonic processes are linked by pyridine nucleotides. This

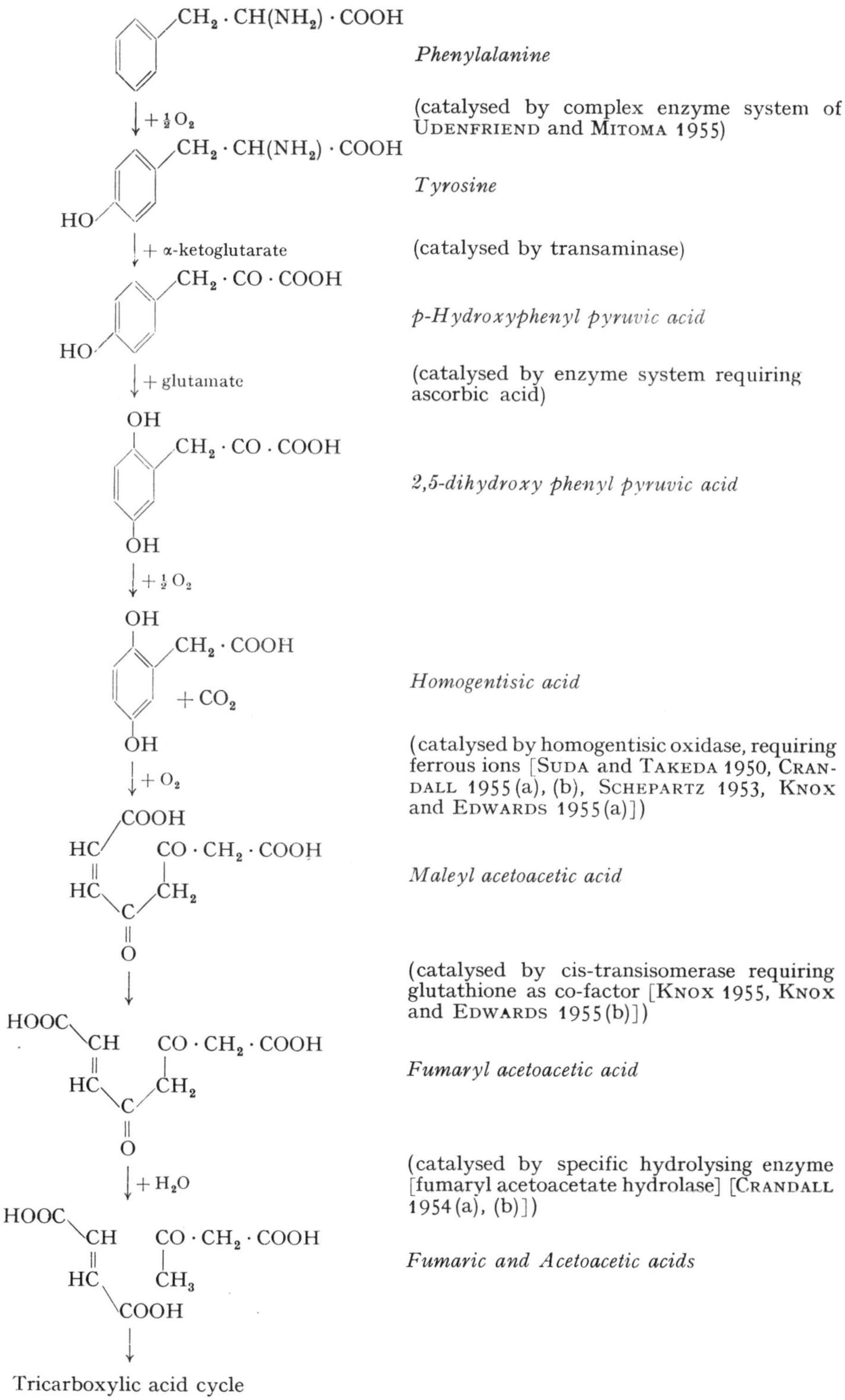

General considerations

$CH_2 \cdot CH(NH_2) \cdot COOH$ — *Phenylalanine*

↓ $+ \frac{1}{2} O_2$ (catalysed by complex enzyme system of UDENFRIEND and MITOMA 1955)

HO—C_6H_4—$CH_2 \cdot CH(NH_2) \cdot COOH$ — *Tyrosine*

↓ + α-ketoglutarate (catalysed by transaminase)

HO—C_6H_4—$CH_2 \cdot CO \cdot COOH$ — *p-Hydroxyphenyl pyruvic acid*

↓ + glutamate (catalysed by enzyme system requiring ascorbic acid)

(OH)$_2 C_6H_3$—$CH_2 \cdot CO \cdot COOH$ — *2,5-dihydroxy phenyl pyruvic acid*

↓ $+ \frac{1}{2} O_2$

(OH)$_2 C_6H_3$—$CH_2 \cdot COOH$ + CO_2 — *Homogentisic acid*

↓ $+ O_2$ (catalysed by homogentisic oxidase, requiring ferrous ions [SUDA and TAKEDA 1950, CRANDALL 1955(a), (b), SCHEPARTZ 1953, KNOX and EDWARDS 1955(a)])

HOOC–CH=CH–C(=O)–CH$_2$–$CO \cdot CH_2 \cdot COOH$ — *Maleyl acetoacetic acid*

↓ (catalysed by cis-transisomerase requiring glutathione as co-factor [KNOX 1955, KNOX and EDWARDS 1955(b)])

HOOC–CH=CH–C(=O)–CH$_2$–$CO \cdot CH_2 \cdot COOH$ — *Fumaryl acetoacetic acid*

↓ $+ H_2O$ (catalysed by specific hydrolysing enzyme [fumaryl acetoacetate hydrolase] [CRANDALL 1954(a), (b)])

HOOC–CH=CH–COOH $CO \cdot CH_2 \cdot COOH$ / CH_3 — *Fumaric and Acetoacetic acids*

↓

Tricarboxylic acid cycle

Fig. 15. *Breakdown of phenylalanine* (KNOX 1955)

represents a short-circuiting of energy transfer. DPN, instead of serving as a substrate in oxidative phosphorylation and thereby generating ATP, reacts directly with the substrate which serves as material for the synthesis. Since such coupled oxido-reductions are generally readily reversible, the efficiency of the energy transfer approaches 100%. Whether a substance is synthesised or degraded in such a reversible system depends on shifts of the equilibrium position brought about by the addition or removal of reactants. As the range of concentrations of reactants is limited, shifts in equilibrium do not usually lead to major net changes. Bulk changes can therefore be brought about only if reversible systems operate in conjunction with systems which are virtually irreversible. Examples of the latter are the hexokinase reaction in glycogen synthesis, the formation of phosphopyruvate from ATP and pyruvate, and dephosphorylation of hexose phosphates in the conversion of lactate to glycogen. In photosynthesis, the carboxydismutase and phosphoribulokinase reactions, and the dephosphorylation of sugar, are the reactions which shift the equilibrium states in one direction. In other words, reversible steps are links connecting the decisive irreversible steps in syntheses. Inasmuch as these irreversible stages usually include reactions in which ATP takes part, it is true to say that ATP is a key fuel which drives the syntheses.

9. Control of Energy-Supplying Processes

The nature and the rates of energy-supplying processes in living cells are not constant but vary with the physiological state — rest or activity — and the environment — the types of available nutrients, the pH and other factors. In cells, mechanisms exist which adjust the energy metabolism according to circumstances. Thus, energy is obtained by oxidations if air is available but by fermentations of sugar if conditions are anaerobic. The rate of energy supply can be increased by cells when they change from the resting to the active state and new enzymes can be developed, particularly in microorganisms, when the chemical composition of the environment changes.

Hormones and the nervous system are responsible for some of the changes in the reaction rates in higher organisms, but control mechanisms also occur in lower forms of life, such as the unicellular ones in which hormones and nerve cells do not occur. These "primitive" (STADIE 1954) control mechanisms are also present in animals; they are in fact the basic systems upon which the action of hormones or the nervous system is super-imposed. Knowledge of intermediary metabolism and of enzyme systems has sufficiently advanced in recent years to prepare the ground for a study of the enzymic mechanisms which operate in the rate control of metabolic processes, and to give tentative answers to the question of how cells adapt their rates of energy supply to changing needs.

The basic unit which determines the rates is the enzyme substrate system. The quantities of enzyme and of substrate together limit the maximum rates which can be reached under any given condition. Such maximum rates are exceptional under physiological conditions, the limiting factor being usually (but not always) the amount of substrate. The fact that intermediate products do not usually accumulate shows that the substrates of the intermediary enzymes are removed as rapidly as they are formed. The average half life of the acids of the tricarboxylic acid cycle in a rapidly respiring tissue is of the order of a few seconds [KREBS 1954(a)]. Thus the amount of enzyme in the tissue is sufficient to deal with the intermediate as soon as the latter arises; in other words, the amount of available substrate is the factor limiting the rate at which the intermediary step proceeds.

This conclusion is borne out by the fact that addition of the intermediates often increases the rate of intermediary processes. The intermediates of the tricarboxylic acid cycle, for example, all stimulate the rate of respiration of suitable tissue preparations (see KREBS 1943). This indicates that the enzymes attacking the added substrates are not used to full capacity when endogenous substrates, such as glucose, are oxidized.

Although, then, the rate of many steps of intermediary metabolism depends mainly on the concentration of the substrates, this is not true for all steps. There are some reactions, small in number by comparison, where the rates depend on factors other than the amount of enzyme or substrates. These are the "pacemakers" of metabolism. They are the reactions on which the study of the control of metabolism must concentrate, the first task being the identification of the pacemakers, the second the analysis of the mechanism by which their rates are controlled.

There is a general principle which may guide the search for pacemakers. As pacemakers are reactions of variable rate, the level of substrate concentration of the pacemakers must vary inversely with the rate: it must increase when the reaction rate decreases. Whether the rise is appreciable may be expected to depend on the equilibrium position of the preceding step. The study of the concentration level of intermediary metabolites, especially the change of steady state concentrations caused by a change of environmental conditions, may therefore provide information on the nature of pacemakers and control mechanisms. LYNEN (1941) and LYNEN and KOENIGSBERGER (1951), for example, found that the change from aerobic to anaerobic conditions in fermenting yeast cells is accompanied by a rise in orthophosphate, which indicates that a reaction involving phosphate is one of the pacemakers of fermentation. Owing to lack of adequate methods of analysis, information on the steady state levels of intermediary metabolites is still very limited (see HOLZER 1953, 1956).

Pacemakers of anaerobic glycolysis. Of the twelve major steps of anaerobic glycolysis (see Fig. 5) two have been assumed by various authors to act as pacemakers of anaerobic glycolysis. The first is the hexokinase reaction (see LePage 1950) which probably initiates all major metabolic reactions of glucose: the anaerobic fermentations, the complete oxidation, the transformations into glycogen, fat, amino acids or other cell constituents (Fig. 16). Whilst the hexokinase reaction (or a reaction very closely connected with it, possibly the penetration of glucose to the site of hexokinase) evidently determines the rate of glucose consumption it cannot control the energy production from glucose because of the variety of pathways open to glucose 6-phosphate.

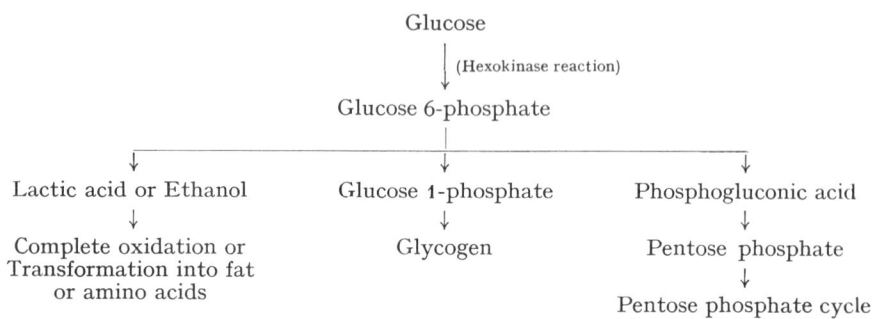

```
                              Glucose
                                 │ (Hexokinase reaction)
                                 ↓
                         Glucose 6-phosphate
                                 │
         ┌───────────────────────┼───────────────────────┐
         ↓                       ↓                       ↓
  Lactic acid or Ethanol   Glucose 1-phosphate    Phosphogluconic acid
         ↓                       ↓                       ↓
  Complete oxidation or      Glycogen              Pentose phosphate
  Transformation into fat                                ↓
      or amino acids                              Pentose phosphate cycle
```

Fig. 16. *Alternative pathways of glucose metabolism*

The triosephosphate dehydrogenase reaction has been taken to be the second pacemaker which decides the rate of energy production from glucose 6-phosphate. As already discussed this is a complex reaction in which DPN,

$$\begin{array}{l} CH_2O \cdot PO_3H_2 \\ | \\ CHOH \\ | \\ HCO \end{array} + ADP + P + DPN \rightleftarrows \begin{array}{l} CH_2O \cdot PO_3H_2 \\ | \\ CHOH \\ | \\ COOH \end{array} + ATP + DPNH_2$$

glyceraldehyde phosphate   *3-phosphoglyceric acid*

Fig. 17. *Over-all change in the triosephosphate dehydrogenase system*

This is the result of two major steps, those catalysed by triosephosphate dehydrogenase and by phosphoglycerokinase. The details of the reaction mechanism are set out in Fig. 2 (Bücher 1947)

ADP and orthophosphate take part (Fig. 17). The detailed chemical change is shown in Fig. 2. It is obvious from the formulation that the reaction cannot take place unless ADP and P are available, — the very two substances which become available when energy is spent. It must be understood, however, that the rate control is due not so much to the presence or absence of these reactants as to the level of their concentrations: the rate decreases already before ADP or P disappear completely. It also follows that the rate of the triosephosphate dehydrogenase reaction may be influenced by any process which either removes or supplies ADP or P. The main reaction which removes ADP and P is respiration, coupled with oxidative phosphorylation. Hence respiration must inhibit fermentation; this is the interpretation of the Pasteur effect (the inhibition of fermentation by oxygen) proposed by Lynen (1941) and JOHNSON (1941). On the other hand any energy requiring process, by supplying ADP or P, will, within limits, stimulate fermentation.

Whilst it is very probable that this mechanism operates, there remain some aspects of glucose utilisation to be clarified, as LYNEN and KOENIGSBERGER (1951) have pointed out. Stoppage of the triosephosphate dehydrogenase reaction can explain stoppage of lactic acid formation, but it does not account for the decreased utilisation of sugar. Non-removal of triosephosphate would lead to an accumulation of fructose diphosphate, because of the ready reversibility of the aldolase reaction (see Fig. 10). But the stage between fructose 1:6-diphosphate and fructose 6-phosphate is not readily reversible. Hence the accumulation of triosephosphate or fructose 1:6-diphosphate cannot by mass action prevent the hexokinase reaction or the conversion of glucose 6-phosphate into fructose 6-phosphate and fructose 1:6-diphosphate. On the contrary, the relatively high concentration of ATP required for the inhibition of the triosephosphate dehydrogenase reaction is expected to facilitate the formation both of glucose 6-phosphate and of fructose 1:6-diphosphate from glucose or from glycogen. Thus additional factors must be responsible for the decrease of sugar consumption when conditions change from anaerobiosis to aerobiosis. LYNEN and KOENIGSBERGER (1951) have raised the questions whether the rate of the hexokinase reaction might be controlled by the hexose phosphates formed. An important observation in this context is the non-competitive inhibition of hexokinase by glucose 6-phosphate, discovered by WEIL-MALHERBE and BONE (1951) and further investigated by CRANE and SOLS (1954). Glucose 6-phosphate in as low a concentration as 0.5×10^{-3} M causes 40% inhibition. It is certainly feasible that this inhibition, or analogous effects at other stages of glycolysis, play a part in the control mechanism, but before this can be accepted, further investigations are needed, especially of the steady state concentrations of the phosphorylated intermediates, of the kinetics of the enzyme systems concerned, and of specific inhibitions by intermediary metabolites.

It has also been considered whether the reduction of sugar consumption in the presence of oxygen might be caused by a direct action of oxygen on an enzyme, for instance by the oxidation of a SH group which might inactivate an enzyme reversibly (LIPMANN 1933, 1934, BURK 1939, MEYERHOF and FIALA 1950, LYNEN and KOENIGSBERGER 1951), but the evidence argues against this kind of mechanism: fermentation can, under a variety of conditions, reach the anaerobic level in the presence of oxygen, as, for example, on addition of dinitrophenol, HCN, or isonitrile.

Pacemakers of respiration. When energy is released by the oxidation of carbohydrate, fat, and amino acids, there are over a hundred identifiable intermediate steps, only a few of which are pacemakers. It follows from the non-accumulation of intermediates that those steps which initiate the oxidation of a substrate must be among the pacemakers of respiration. These reactions also decide which substrate, among a mixture, is attacked preferentially — whether carbohydrate, fatty acids or amino acids serve as a source of energy.

In addition, pacemakers are expected at two other stages of the oxidative metabolism, at those where the total oxygen consumption (i.e. energy supply) is determined and at those where, after a partial degradation, more than one pathway is open. A diagram of the fate of carbohydrate and fatty acids (Fig. 18) shows the major stages where branching of pathways may occur. It is decided at these branching points whether glucose is broken down to supply energy or stored in the form of glycogen; whether glucose, via acetyl

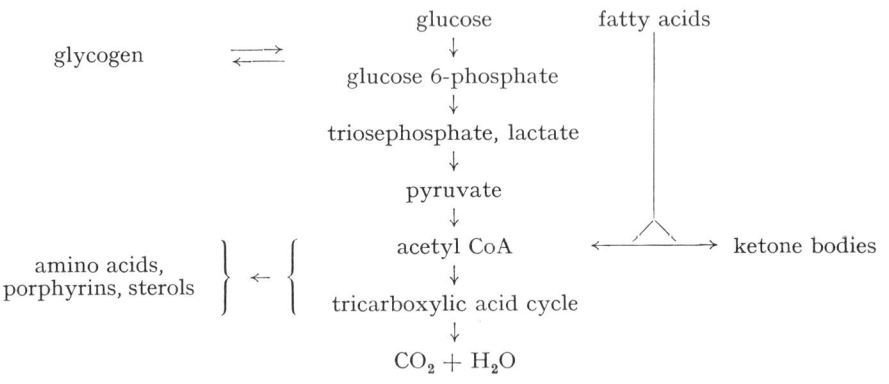

```
                              glucose          fatty acids
                                 ↓                 |
      glycogen    ⇌     glucose 6-phosphate        |
                                 ↓                 |
                       triosephosphate, lactate    |
                                 ↓                 |
                              pyruvate             |
                                 ↓                 |
      amino acids,             acetyl CoA   ←——→  ketone bodies
      porphyrins, sterols  } ← {    ↓
                          tricarboxylic acid cycle
                                 ↓
                             $CO_2 + H_2O$
```

Fig. 18. *Branching of metabolic pathways*

coenzyme A, is converted into fat or whether fat is broken down to give energy; whether ketone bodies are formed from fatty acids or acetyl coenzyme A (ketogenesis) or disposed of via the tricarboxylic acid cycle (antiketogenesis); whether intermediates of the tricarboxylic acid cycle are oxidized, or used to supply carbon skeletons for the synthesis of amino acids, porphyrins or steroids. Two of the three stages where pacemakers are expected — the initiation of the breakdown and the branching points — have a feature in common: there is in both cases a choice between alternative reactions, in the first case between different substrates, and the second between different routes starting from the same substrate.

Enzymic pattern of control mechanism. The analysis of the control mechanisms which operate in these pacemaker reactions must begin with a translation of the situation into the terminology of enzyme chemistry. Two types of situations may be distinguished. Alternative metabolic routes may arise either from a choice of alternative *reactions* (say, breakdown of either carbohydrate, or fat, or protein, when energy is needed), or a choice of alternative directions of pathways (say, synthesis or breakdown of fatty acids). In the first case the reactions which take place are essentially different, whilst in the second the same reactions move in opposite directions.

Control of Energy-Supplying Processes

The closer examination of the first case shows that the alternative reactions which occur when cells have the choice of obtaining energy from a variety of substrates are not entirely independent and separate processes. As already discussed in Chapters 2 and 3, only the first step of the complex sequence by which the substrate is brought to reaction with molecular oxygen varies from substrate to substrate. All other intermediate steps [reactions (3, 2) to (3, 5)] are shared by the alternative reactions as diagrammatically shown in Fig. 19. Five substrates representing the great majority of metabolites supplying energy

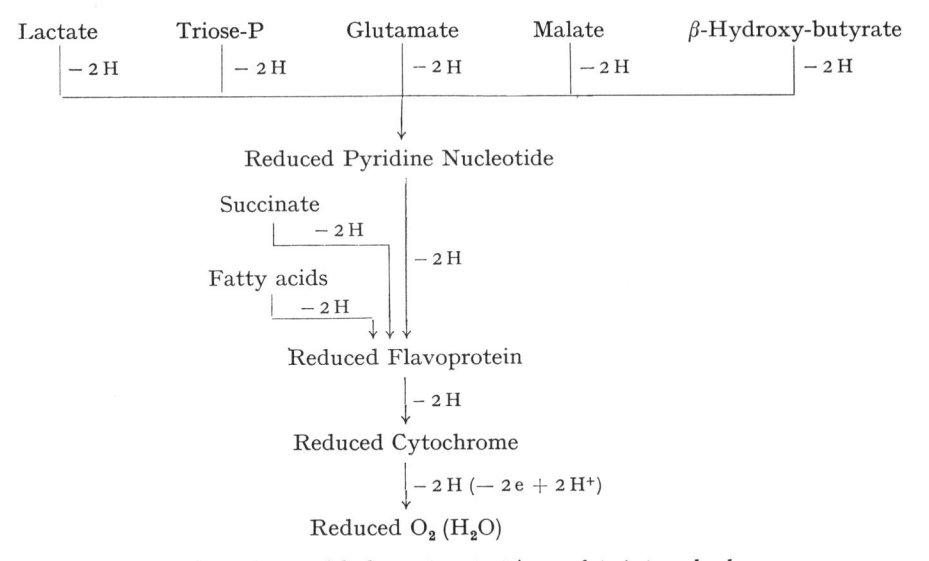

```
 Lactate     Triose-P     Glutamate      Malate      β-Hydroxy-butyrate
  | −2 H      | −2 H       | −2 H         | −2 H        | −2 H
  └────────────┴─────────────┬─────────────┴──────────────┘
                             ↓
                  Reduced Pyridine Nucleotide
              Succinate       |
               | −2 H         |
                   ┐       −2 H
          Fatty acids        |
               | −2 H        |
                   ↓ ↓ ↓
                  Reduced Flavoprotein
                       | −2 H
                       ↓
                  Reduced Cytochrome
                       | −2 H (−2 e + 2 H⁺)
                       ↓
                  Reduced O₂ (H₂O)
```

Fig. 19. *Joint pathways of hydrogen transport from substrate to molecular oxygen*

are chosen to illustrate the principle. The oxidation begins in every case with a transfer of hydrogens atoms to a common reagent, DPN. Partial exceptions, as already mentioned in Chapter 3, are fatty acids and succinate, but most intermediate steps are in these cases also shared with the other substrates. Fatty acids and succinate donate hydrogen atoms directly to flavoprotein, deviations from the rule which are necessitated by the thermodynamic characteristics of these substrates.

Choice of substrate thus means competition of several substances for the same catalyst. The majority of substrates compete directly for DPN. Fatty acids, succinate and reduced DPN compete for flavoprotein. Which substrate is oxidized depends on physico-chemical properties which are definable, such as the relative rates at which the substrates interact with the common agent.

The pattern of the enzymic mechanisms is somewhat different when the alternatives are different directions of the same reactions, e.g. synthesis or breakdown of fatty acids. The control mechanism for such a system has already been discussed in Chapter 8. In this case the relative values of the ratios

$\frac{\text{acetyl CoA}}{\text{CoA—SH}}$, $\frac{\text{reduced flavoprotein}}{\text{oxidized flavoprotein}}$, and $\frac{\text{DPNH}_2}{\text{DPN}}$ determine the direction of the reactions by mass action and the value of these ratios in turn depends above all on the availability of pyruvate.

There are many other syntheses where the mechanisms controlling the balance between synthesis and breakdown cannot yet be satisfactorily visualised, mainly because the enzymic reactions have not been sufficiently clarified. The principles operating in the fatty acid system are probably factors in many cases. An additional factor is the presence of special enzymes such as those which overcome the energy barriers in the synthesis of carbohydrate from lactate or pyruvate (Chapter 8).

The mechanism which regulates the rate of oxygen consumption, i.e. the rate at which $DPNH_2$, reduced flavoprotein and reduced cytochrome c react with O_2, is of a different type. As already discussed, these reactions are coupled with the synthesis of ATP from ADP and orthophosphate. Owing to this coupling the reaction

$$DPNH_2 + \tfrac{1}{2}O_2 \to DPN + H_2O \qquad (3, 5)$$

and its component steps (3, 2), (3, 3) and (3, 4) proceed at full rates only if ADP and orthophosphate are present, [although short cuts not involving phosphorylation may possibly occur (LEHNINGER 1954)]. The role of phosphates in this reaction is analogous to that in the triosephosphate dehydrogenase reaction. Both reactions are controlled by the rate at which ATP is split to ADP and P, i.e. by the amount of energy spent. As the rate of the interaction between substrates and DPN depends on the rate at which DPN is regenerated by the oxidation of $DPNH_2$, oxidative phosphorylation also controls the rate of degradation of the substrate.

These concepts, it should be repeated, are tentative. They are supported by substantial evidence but cannot be regarded as established in detail. What is well established is the rate-limiting function of P and/or phosphate acceptors under certain well defined conditions. Earlier observations by LENNERSTRAND (1936) and BELITZER (1939) on rate control by phosphate and/or phosphate acceptors have been elaborated in particular by LARDY and WELLMAN (1952), who showed that the rate of oxidation of a variety of substrates in liver mitochondria is greatly accelerated by the addition of inorganic phosphate and of ADP or other phosphate acceptors (see LARDY 1952, COOPER, DEVLIN and LEHNINGER 1955, CHANCE 1955). RABINOVITZ, STULBERG and BOYER (1951) obtained similar results with heart muscle. Under some experimental conditions the rate-limiting factor is inorganic phosphate, under others it is the phosphate acceptor. It is probable that both can play a role *in vivo*.

Contol of ketogenesis. Acetoacetate, the chief of the ketone bodies, arises mainly by condensation of two molecules of acetate which react in the form of acetyl coenzyme A. As already stated, acetyl coenzyme A has the alternative of entering the tricarboxylic acid cycle at one of the branching points. As this route requires oxaloacetate, this substance is taken to be the key metabolite in ketogenesis and antiketogenesis (KORÁNYI and SZENT-GYÖRGYI 1937). However, experimental tests in the intact organisms failed in many cases to bear out this view: succinate, which readily supplies oxaloacetate in the body, does not relieve acidosis in human diabetics (LAWRENCE 1937, LAWRENCE, MCCANCE and ARCHER 1934) and in the rat (TERRELL 1938, DEUEL, HALLMAN and MURRAY 1937), though BEATTY and WEST (1951) noted some decrease in urinary ketone body excretion in rats made ketotic by excess butyrate, when various precursors of oxaloacetate were administered by stomach tube. On the other hand, oxaloacetate is certainly antiketogenic in isolated mitochondria. Whilst fatty acids yield ketone bodies in liver mitochondria in the absence of oxaloacetate, they are oxidized through the tricarboxylic acid cycle when oxaloacetate is added (LEHNINGER 1946). The availability of oxaloacetate can thus, at least under certain conditions, decide between ketone body accumulation and complete oxidation of fatty acids.

Experiments with substances which interfere with the tricarboxylic acid cycle and thereby reduce the supply of oxaloacetate lead to the same conclusion. Malonate, which prevents the oxidation of succinate, and ammonium ions, which divert α-ketoglutarate to glutamate (RECKNAGEL and POTTER 1951) are both ketogenic. These are observations which point to the key role of oxaloacetate, but it cannot be claimed that the mechanism controlling ketogenesis is fully understood, probably because of insufficient information on the factors which control the steady state level of oxaloacetate. At least five independent reactions are known to cause the formation or disappearance of oxaloacetate:

$$\text{malate} + \tfrac{1}{2} O_2 \rightleftarrows \text{oxaloacetate} + H_2O \text{ (malic dehydrogenase)}$$

$$\text{phosphopyruvate} + CO_2 + \text{ADP (or IDP)} \rightleftarrows \text{oxaloacetate} + \text{ATP (or ITP)} \text{ (UTTER-KURAHASHI reaction)}$$

$$\text{aspartate} + \alpha\text{-ketoglutarate} \rightleftarrows \text{oxaloacetate} + \text{glutamate (transaminase)}$$

$$\text{oxaloacetate} + H_2O \rightleftarrows \text{pyruvate} + HCO_3^- \text{ (oxaloacetic decarboxylase)}$$

$$\text{oxaloacetate} + \text{acetyl coenzyme A} \rightleftarrows \text{citrate} + \text{coenzyme A (condensing enzyme)}$$

The steady state level of oxaloacetate must depend on the interplay of these five (and possibly other) reactions. How this interplay is governed is unknown.

In summing up this aspect of ketogenesis it can be said that the metabolic fate of acetyl coenzyme A at one of the branching points is decided by the supply of a special reactant. Again there is competition of one substrate between two alternatives, the alternative in this case being either a reaction with oxaloacetate or a reaction with the second molecule of acetyl coenzyme A.

The availability of oxaloacetate can account for the formation or nonformation of acetoacetyl coenzyme A. Free acetoacetate appears only in liver in appreciable quantities. This is accounted for by the fact that liver contains an enzyme which hydrolyses acetoacetyl coenzyme A ('deacylase') and, unlike most other tissues, cannot convert free acetoacetate back to the coenzyme A derivative (MII and GREEN 1954, MAHLER 1953). This again illustrates the obvious point that presence or absence of specific enzymes contributes to the control of pathways.

General comments on the design of control mechanisms

Hormonal control of biological activities operates generally through an effect of the hormone on the activity of an enzyme. This may be accelerating or inhibitory, probably more often inhibitory. The non-hormonal control mechanisms discussed here are in principle of a different type. They usually belong to the type of systems which engineers call "feedback"systems. Control by feedback is an arrangement in which the controlled process, as it progresses, creates conditions unfavourable for further progress, and thereby causes the rate to decrease. This slowing down recreates more favourable conditions and thereby speeds up the process. Thermoregulators are among the simplest examples of feedback systems.

The mechanisms by which biological energy transformations are controlled include in many cases arrangements of the same type. The key substances through which these controls are operated appear to be inorganic phosphate and ADP. Their presence stimulates the rate of both the aerobic and the anaerobic degradation of foodstuffs. Their concentrations are bound to rise when energy is spent. The increased rate of respiration or fermentation in turn causes their removal and thereby decreases the rates.

Competitive mechanisms, in which DPN is one of the key substances, may also be looked upon as feedback systems. When two substrates compete for a common intermediary catalyst, each, by its presence, creates unfavourable conditions for the reaction of the other substrate. As one of the substrates disappears, the second is automatically fed to the catalyst so that approximate constancy of catalytic activity is secured.

The relative contributions made in higher animals by hormonal control and feedback control may be illustrated by the example of the energy-supplying reactions. Whilst hormones, in particular that of the thyroid, play an important role in the control of the basal metabolic rate, the energy needs due to functional activity, such as contraction by muscles, or secretion by glands, seem to be controlled by feedback mechanisms. It is true that hormones may control secretion and thereby influence the rate of energy supply. But these effects of hormones on the rates of energy supply are probably indirect and achieved through the intermediation of feedback mechanisms.

Feedback mechanisms occur at many other levels of biological organisation. The long known effect of the alveolar CO_2 pressure on breathing (HALDANE and PRIESTLEY 1905) is an example. Others concern the activities of the nervous system, including those of the higher centres. The purposeful behaviour of biological systems in particular can in many cases be accounted for in an entirely mechanistic manner by hypotheses based on the feedback principle (ROSENBLUETH, WIENER and BIGELOW 1943).

So far, no more than a beginning has been made in the elucidation of the chemical mechanisms which control the nature and the rate of energy-supplying processes, but an insight has already been gained into some of the principles involved. The adaptation of the energy-supply to changing needs can, at least in part, be understood on the basis of the properties of the enzyme systems, inhibitions and feedback arrangements being the distinctive features of the control mechanisms.

10. A Special Feature of ATP as an Energy Store

Energy-rich phosphate bonds of ATP can be synthesised at the expense of the free energy of many types of oxido-reductions, irrespective of the level of the redox potential of the systems concerned. As long as the two interacting systems have suitable concentrations and potentials to yield sufficient free energy, they can support the synthesis of ATP. For example, in oxidative phosphorylation, ATP may be formed by the oxidation of $DPNH_2$ to DPN (standard redox potential -0.320 volt at pH 7.0), by the oxidation of dihydroflavoprotein (standard redox potential about -0.06 volt at pH 7.0) and by the oxidation of ferrocytochrome c (standard redox potential $+0.26$ volt at pH 7.0).

The fact that the synthesis of, and the release of energy from, ATP is independent of the redox scale allows ATP to be used as an "energy currency" or "energy transmitter" in a wide range of situations. It can in particular act as an energy link between two oxido-reduction systems. This means that one oxido-reduction (e.g. the oxidation of ferrocytochrome by O_2), by producing ATP, can theoretically drive another similarly ATP-producing oxido-reduction backwards. Thus the system $DPNH_2$ + flavoprotein $\rightleftarrows$ DPN + dihydroflavoprotein could be driven from right to left, to produce $DPNH_2$, provided that the coupling of oxido-reduction and ATP synthesis in this system is reversible (DAVIES & KREBS 1952). Whether this theoretical possibility is in fact realised has not been definitely established but experimental observations strongly suggest that it occurs. As has been previously mentioned the reductive synthesis of malate from pyruvate and CO_2, catalysed by the malic enzyme:

$$\text{pyruvate} + CO_2 + TPNH_2 \rightarrow \text{malate} + TPN \qquad (8, 12)$$

is much more rapid in the presence of O_2 than in its absence (KREBS 1954). So far the only satisfactory explanation for this effect is the assumption that

O_2, through the intermediation of ATP, facilitates the reduction of TPN to $TPNH_2$. This assumption is supported by the observation that dinitrophenol abolishes the effect of O_2. It is at first surprising that O_2 can promote a reduction in this system despite the fact that oxygen, if in direct contact with a suitable catalyst, causes the oxidation of $TPNH_2$. The situation regarding the action of ATP is exactly analogous to the action of electrons in a system of oxidation-reduction cells which allow one oxido-reduction cell to drive another cell backwards, even though the components of this cell may have lower oxido-reduction potentials and would be oxidized by the components of the first cell if they were mixed.

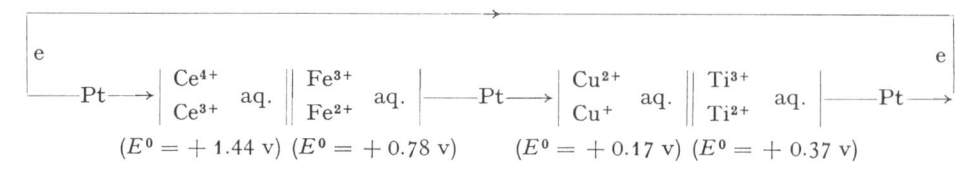

$$e \quad\quad Pt \to \left|\begin{array}{c}Ce^{4+}\\Ce^{3+}\end{array}\text{aq.}\right|\left|\begin{array}{c}Fe^{3+}\\Fe^{2+}\end{array}\text{aq.}\right| \quad Pt \to \left|\begin{array}{c}Cu^{2+}\\Cu^{+}\end{array}\text{aq.}\right|\left|\begin{array}{c}Ti^{3+}\\Ti^{2+}\end{array}\text{aq.}\right| \quad Pt \to \quad e$$

$$(E^0 = +1.44\text{ v}) \quad (E^0 = +0.78\text{ v}) \quad\quad (E^0 = +0.17\text{ v}) \quad (E^0 = +0.37\text{ v})$$

Fig. 20. *Oxidation-reduction cells coupled by wire*
(For particulars see text; DAVIES and KREBS 1952)

In the electric cells shown in Fig. 20, the first cell contains a solution of Ce^{4+} and Ce^{3+} salts in one compartment and of Fe^{3+} and Fe^{2+} salts in another. The net E. M. F. of this cell is $1.44 - 0.78 = 0.66$ v, and the generation of current is accomplished by the reaction

(a) $Ce^{4+} + Fe^{2+} \to Ce^{3+} + Fe^{3+}$.

The second cell contains solutions of Cu^+ and Cu^{2+} in one compartment and of Ti^{3+} and Ti^{2+} in another. The net E. M. F. of this cell is $0.37 - 0.17 = 0.20$ v, and the generation of current is accomplished by the reaction

(b) $Ti^{3+} + Cu^+ \to Ti^{2+} + Cu^{2+}$.

When the two cells are connected as shown in the diagram, the greater E. M. F. of the first cell can drive the second cell backwards, and the reaction now occurring in the second cell is the reverse of (b), i.e.

(c) $Cu^{2+} + Ti^{2+} \to Cu^+ + Ti^{3+}$.

The net effect is, therefore, that the reduction of Cu^{2+} is brought about by the powerful oxidiser, Ce^{4+}.

It is no doubt of great significances that in a system which derives its energy in the last resort from oxido-reductions, the free energy of the hydrolysis of ATP is independent of redox potentials. Like the electrons in the platinum wire, ATP can act as "energy currency" and link energetically systems of widely different redox potentials.

11. Evolution of Energy Transforming Mechanisms

The object of this survey, as stated at the outset, has been to discuss those aspects of biological energy transformations which are common to many different types of organisms. These common features reveal a continuity between chemical processes in different types of organisms, and also between different types of chemical processes within the same organism, and thus lead to a unifying concept: many types of chemical activities which at first may appear totally different are in fact modifications of the same basic themes.

It has been argued that the emphasis on unifying principles of this sort is artificial and misleading. In reply it may be said that such concepts are the very essence of evolutionary theory. This theory envisages a continuity between all types of living organisms. The manifestation of this continuity is a sharing of biological properties, which occurs not only at the level of energy transformations (as stressed in this survey) but at all levels of biological organisation. At the morphological level it is, for example, the cell and its differentiation into cytoplasm, nucleus and other organelles which are shared by all forms of life (except the incomplete parasitic ones of viruses). At the physiological level, the mechanisms which effect exchanges of materials between cells and their environment — absorption and secretion — may be cited as universal biological phenomena.

The shared basic themes must be taken as representing the starting levels from which numerous additional activities have evolved. Thus the full picture of energy transformations contains a much greater variety of metabolic processes than has been shown in the present account.

The newer knowledge of the common features of the different metabolic processes has contributed further evidence to the hypotheses about biochemical evolution advanced in recent years (HALDANE 1929, OPARIN 1938, HOROWITZ 1945, UREY 1952). Reactions which show the widest occurrence among living organisms are those taking part in the anaerobic fermentation of hexoses to lactic acid or ethanol. This suggests that these fermentations arose very early in the evolutionary scale (HALDANE 1929). It appears to be generally accepted that, when life came into being, the earth's atmosphere was a reducing one, and that organic compounds were plentiful (HALDANE 1929, 1954, OPARIN 1938, DAUVILLIER and DESGUIN 1942, HOROWITZ 1945, VAN NIEL 1949, UREY 1952, BERNAL 1954, PIRIE 1954). Evidence to support this view largely rests on geochemical grounds (UREY 1952) and on the demonstration that organic materials found in biological systems, such as some organic acids and amino acids, can be synthesised by the passage of high-energy radiation through a reducing atmosphere of methane, ammonia, hydrogen and water MILLER 1953, 1955).

The view that anaerobic fermentations were the earliest energy-supplying processes is consistent with the finding that some of the components of the fermentation reactions, including ATP and DPN, occur in almost every other metabolic process supplying, or depending on, utilisable energy (see BLUM 1951, MADISON 1953). It is further noteworthy that although ATP and DPN are functionally very different, they are chemically closely related, in so far as DPN contains the adenylic acid moiety.

There is no clear-cut evidence to indicate the time in the evolutionary scale when the reactions of the pentose phosphate cycle arose. It is likely that the earliest organisms did not require this metabolic mechanism. Pentoses may have been abundant in the primeval environment, and the pentose phosphate cycle, which supplies pentoses to the organism, probably evolved only when the original environmental pentose supply decreased. Further evidence to support the view that this cycle evolved after the fermentations is provided by the fact that the cycle contains several of the reactions of fermentation as well as some new ones. The close similarity of the reactions of the pentose phosphate cycle and the path of carbon in photosynthesis raises the question which of the two mechanisms arose first. Some authors (OPARIN 1938, UREY 1952, MILLER 1953, 1955, but cf. GOLDSCHMIDT 1952) assume that free carbon dioxide was almost absent from the earth's atmosphere when life began. If this was the case, photosynthesis must have arisen after carbon dioxide had been released, both by biological processes and by possibly its liberation from carbonate rocks. The concept that the complexity of biological mechanisms has increased gradually suggests that the pentose phosphate cycle arose before photosynthesis. Given the reactions of fermentation and of the pentose phosphate cycle, only a few additional reactions are required for photosynthesis, namely two reactions in the pathway of carbon and the complex photochemical reaction proper.

OPARIN (1938), UREY (1952) and others have concluded that the oxygen of the atmosphere was essentially produced by photosynthesis. The appearance of photosynthesis would thus pave the way for the evolution of respiratory mechanisms. The most widespread reactions for the oxidative utilisation of food — the tricarboxylic acid cycle operating in conjunction with the cytochrome system — would therefore have arisen later in the evolutionary sequence than the fermentations and the pentose phosphate and photosynthetic carbon cycles.

These tentative time relationships in the evolution of the main biological energy-transforming processes may be visualised as shown in Fig. 21. In view of its widespread occurrence amongst all types of aerobic organisms, even the last stage shown in this diagram, that of cell respiration, must have evolved comparatively early, at a stage when organisms were still unicellular.

Discussions have taken place from time to time about the 'unity' or 'disunity' of the biochemical organisation of different types of organism. Some authors consider it as more profitable to emphasise different rather than common features. Thus COHEN (1954) suggests that comparative biochemistry should concern itself with "the study of the origin, nature and control of biochemical variability". On the other hand, STANIER (1954), following KLUYVER (1931) and VAN NIEL (1949) defines the task of the comparative biochemist as "to seek for the common biochemical principles which are expressed in all forms of life". These differences merely reflect different attitudes towards the same

Anaerobic fermentations (include glycolytic enzymes, ATP and pyridine nucleotide)
↓
Pentose phosphate cycle
↓
Photosynthesis
(includes metalloporphyrins)
↓
Cell respiration
(includes tricarboxylic acid cycle, oxidative phosphorylation, cytochrome system)

Fig. 21. *Tentative evolutionary time scale of the main reactions concerned with energy transformations*
(The arrows indicate time intervals. The scale is based on the assumption that the surface of the primitive earth was rich in organic substances, but free from O_2 and CO_2 and further that the later stages have evolved from older ones by the implantation of new reactions upon existing mechanisms. For further details see text)

facts. Some authors prefer to lay stress on endless variety and complexity, and others to trace the common ancestry from which the variety has sprung, but most will agree on the need for both approaches.

Appendix
Free Energy Data of Biological Interest

By

K. BURTON

Free energy data indicate the approximate equilibrium positions of reactions where the equilibrium concentrations of reactants have not been measured directly. In addition they are of more general use in studying energy transformations as indicated in the preceding review. It must be remembered, however, that the difficulties of evaluating the concentrations and thermodynamic activities of substances under the complex conditions existing in cells often impose severe limitations on the use of free energy data.

The following tables summarise most of the reliable free energy data at present available for compounds and reactions of biological interest. Some values have been newly calculated; others, where necessary, have been revised in the light of changes in the values accepted for the ancillary data used in

Table 9. *Free energies of formation from the elements*

Name	$-\Delta G°f$ (kgcal) Pure compound	$-\Delta G°f$ (kgcal) In aqueous solution	References and comments
Acetaldehyde (gas)	31.96	33.38	Rossini et al. (1952); cf. Burton and Wilson (1953). Solution data from Wurmser and Wurmser (1936).
Acetic acid (liq.)	93.75	95.48	Burton and Krebs (1953). Uncertainty (± 1 kgcal) in combustion data.
Acetate$^-$	—	88.99	
Acetoacetate$^-$	—	118	Approximate value from β-hydroxybutyrate and redox potential (see Table 10).
Acetone (liq.)	37.18	38.52	Burton and Wilson (1953).
cis-Aconitate^{3-}	—	220.51	Burton (1955).
Adenine	−71.55	—	Stiehler and Huffman (1935).
L-Alanine	88.40	88.75	Burton and Krebs (1953).
Allantoin	106.62	—	Stiehler and Huffman (1935).
NH$_3$ (gas)	3.98	6.37	Rossini et al. (1952).
NH$_4^+$	—	19.00	
L-Arginine	57.44	—	Huffman and Ellis (1937).
L-Asparagine	183.50 (monohydr.)	125.86	Borsook and Huffman (1938). Value adjusted slightly in view of data for anh. compound and for heats of solution.
L-Aspartic acid	174.76	172.31	Burton and Krebs (1953). Combustion data do not agree with those of Oka [1944(a)].
L-Aspartate^{+2-}	—	166.99	
n-Butanol (liq.)	40.39	41.07	Burton and Krebs (1953).
n-Butyric acid (liq.)	90.65	90.86	Burton and Krebs (1953), Hansen, Miller, and Christian (1955). No Washburn correction.
n-Butyrate$^-$	—	84.28	
CO$_2$ (gas)	94.26	92.31	Rossini et al. (1952).
HCO$_3^-$	—	140.31	
Citrate^{3-}	—	279.24	Burton (1955).
iso-Citrate^{3-}	—	277.65	
Creatine	63.19 (anh.)	—	Borsook and Huffman (1938).
	121.11 (monohydr.)	63.17	
Creatinine	6.90	6.91	
Cysteine	163.55	159.00	Combustion data not accurate; no Washburn correction (cf. Skinner 1954). Values calculated from Borsook, Ellis and Huffman (1937) and adjusted to be consistent with Kolthoff, Stricks and Tanaka (1955).
Cystine	82.08	81.21	

Table 9. (Continued)

Name	$-\Delta G°f$ (kgcal) Pure compound	$-\Delta G°f$ (kgcal) In aqueous solution	References and comments
Dulcitol	227.19	—	PARKS, KELLEY and HUFFMAN (1926), PARKS, WEST, NAYLOR, FUJII and MCCLAINE (1946).
Erythritol	154.09	—	
Ethanol (liq.)	41.77	43.39	ROSSINI et al. (1952), BUTLER, RAMCHANDANI and THOMSON (1935).
Ethylene glycol (liq.)	77.12	—	ROSSINI et al. (1952).
Formaldehyde (gas)	26.30	31.2	ROSSINI et al. (1952), PARKS and HUFFMAN (1932).
Formic acid (liq.)	82.7	85.1	ROSSINI et al. (1952).
Formate⁻	—	80.0	
		83.77	WOODS (1936). Apparent inconsistency.
Fructose	—	218.78	Calculated from glucose and epimerization data (WOLFROM and LEWIS 1928). Agrees with assumed entropy of 52.8 E.U. and with heat of combustion (CLARKE and STEGEMAN 1939).
Fumaric acid	156.49	154.67	BURTON and KREBS (1953).
Fumarate²⁻	—	144.41	
α-D-Galactose	220.00	220.73	CLARKE and STEGEMAN (1939), JACK and STEGEMAN (1941), TALLEY and HUNTER (1952). Activity coefficient of 1.28 assumed for saturated solution.
α-D-Glucose	217.56	219.22	HUFFMAN and FOX (1938) TAYLOR and ROWLINSON (1955).
L-Glutamic acid	173.81	171.76	BORSOOK and HUFFMAN (1938). Combustion and entropy data confirmed by OKA [1944 ( ), (b)].
L-Glutamate⁺²⁻	—	165.87	
Glycerol (liq.)	114.02	116.76	BURTON and KREBS (1953).
Glycine	88.61	89.26	ROSSINI et al. (1952), BORSOOK and HUFFMAN (1938).
Glycogen	—	158.3	BURTON and KREBS (1953). Value is for one glucose unit. Independent of concentration.
Glycollate⁻	—	126.9	From glyoxylate and redox potential (Table 10).
Glyoxylate⁻	—	112.0	Approximate value from METZLER, OLIVARD and SNELL (1954) and appropriate $\Delta G°f$ values.
Guanine	−11.23	—	STIEHLER and HUFFMAN (1935).
H₂O (liq.)	56.69	—	ROSSINI et. al. (1952).
OH⁻	—	37.60	

Table 9. (Continued)

Name	−ΔG°f (kgcal) Pure compound	−ΔG°f (kgcal) In aqueous solution	References and comments
H_2O_2	—	32.67	Lewis and Randall (1923). Re-calculated using data for BaO and Ba^{++} from Rossini et al. (1952). Approximate.
H_2S (gas)	7.89	6.54	Rossini et al. (1952).
HS$^-$	—	−3.0	Kury, Zielen and Latimer (1953).
β-Hydroxybutyric acid	—	127	Approximate values by analogy from butyric acid cf. Burton and Krebs (1953).
β-Hydroxybutyrate$^-$	—	121	
Hypoxanthine	−18.39	−21.4	Stiehler and Huffman (1935), Albert and Brown (1954).
α-Ketoglutarate	—	190.62	Burton and Krebs (1953).
Lactate$^-$	—	123.76	Burton and Krebs (1953).
α-Lactose	419.11 (monohydr.)	362.15	Burton and Krebs (1953). Data are inconsistent. N.B. Heats of combustion are not consistent with each other and with heats of solution (Hudson and Brown 1908).
β-Lactose	374.56 (anh.)	375.26	
L-Leucine	82.63	81.68	Borsook and Huffman (1938). Value adjusted (+0.03 kgcal) to take account of those for D- and DL-leucine and ΔG⁰ of racemization.
Mannitol	225.20	225.29	Parks et al. (1926, 1946), Findlay (1902).
Malate^{2-}	—	201.98	Burton and Krebs (1953).
β-Maltose	413.48 (monohydr.)	357.80	Activity coefficient of 2 assumed for saturated solution. Not consistent with other data, Burton and Krebs (1953).
Methane (gas)	12.14	—	Rossini et al. (1952).
Methanol (liq.)	39.73	41.88	Rossini et al. (1952).
NO_3^-	—	26.41	Rossini et al. (1952).
NO_2^-	—	8.25	Latimer (1952).
Oxalic acid	166.8	—	Rossini et al. (1952).
Oxalate^{2-}	—	161.3	
Oxaloacetate^{2-}	—	190.53	Burton and Krebs (1953).
Palmitic acid	25.7	—	Parks and Huffmann (1932). No Washburn correction.
n-Propanol	41.21	42.02	Parks et al. (1929), Rossini (1934), Butler et al. (1935).
iso-Propanol	43.26	44.44	Burton and Wilson (1953).
Pyruvate$^-$	—	113.44	Burton and Krebs (1953).
Sorbose	217.57	—	Clarke and Stegeman (1939), Jack and Stegeman (1941).
Sorbitol	—	225.31	From fructose and Table 10.

Table 9. (Continued)

Name	$-\Delta G°_f$ (kgcal) Pure compound	$-\Delta G°_f$ (kgcal) In aqueous solution	References and comments
Succinic acid	178.68	178.39	BURTON and KREBS (1953).
Succinate^{2-}	—	164.97	
Sucrose	369.20	370.90	BURTON and KREBS (1953). Not consistent with other data.
SO_4^{2-}	—	177.34	ROSSINI et al. (1952).
SO_3^{2-}	—	118.8	
$S_2O_3^{2-}$	—	122.7	MEL (1953).
L-Threonine	—	123.0	Approximate. From METZLER, LONGENECKER and SNELL (1954) and data for glycine and acetaldehyde.
L-Tyrosine	96.10	92.55	HUFFMAN and ELLIS (1937), BORSOOK and HUFFMAN (1938). Combustion data do not agree with those of OKA [1944(a)].
Urea	47.12	48.72	ROSSINI et al. (1952).
Uric acid	90.41	85.3	STIEHLER and HUFFMAN (1935), ALBERT and BROWN (1954).
Urate$^-$	—	77.9	
	—	77.8	GREEN (1934) and value for hypoxanthine.
Water (liq.)	56.69	—	ROSSINI et al. (1952).
Xanthine	39.64	—	STIEHLER and HUFFMAN (1935).
	—	33.3	GREEN (1934) and value for hypoxanthine or urate$^-$ ion.

previous calculations. For the sake of space and convenience, the original measurements, on which the free energy values are based, are often not directly cited by references to the literature but sufficient references are given to indicate both the original measurements and the methods of calculation used to obtain the present values.

Although the first decimal place is not significant in most of the free energy values, two decimal places are retained for calculation purposes so that, for example, the free energy of solution of acetaldehyde may be obtained from the difference between the values for the formation of the gaseous and the aqueous compounds. The symbols used here are based on those given in the Report of the Royal Society Commitee on Symbols, 1951 (cf. BURTON and KREBS 1953). ΔG^0 (ΔF^0 of LEWIS and RANDALL 1923) is the increment of free energy under standard conditions, which are 25°C and a pressure of one atmosphere. In aqueous solution, the standard condition of all solutes is 1 molal activity; that of water is the pure liquid. $\Delta G'$ is identical with ΔG^0 except that the standard condition of H$^+$ ion is that of the p_H specified (usually 7) instead of 1 molal activity (p_H 0).

Table 10. *Free energies and potentials of oxidation-reduction reactions in aqueous solution*
Values are for aqueous solution except that CO_2 and O_2 are gases at 1 atmosphere

Reactants	Products (+ H_2 gas)	$\Delta G'$ (kcal) at p_H 7	Eo' (volts) at p_H 7	References and comments
Acetaldehyde + H_2O	Acetate⁻ + H⁺	−8.47	−0.598	From Table 9.
Acetaldehyde + CoA	Acetyl CoA	0.11	−0.412	Burton and Wilson (1953), Burton and Stadtman (1953).
Alanine + H_2O	Pyruvate + NH_4^+	13.0	−0.132	From Table 9.
NH_4^+ + H_2O	$NH_2OH \cdot H^+$	45.0	0.562	Rossini et al. (1952).
Ascorbate⁻ + H⁺ (p_H 4)	Dehydroascorbic acid	—	0.166 (p_H 4)	Borsook, Davenport, Jeffreys and Warner (1937).
Aspartate⁻ + H_2O	Oxaloacetate²⁻ + NH_4^+	14.15	−0.107	From Table 9.
2:3 Butylene glycol	Acetyl methyl carbinol	7.83	−0.244	Burton and Wilson (1953), Strecker and Harary (1954).
Butyryl CoA	Crotonyl CoA	27.7	0.187	Green, Mii, Mahler and Bock (1954). Equilibria approached from one direction only.
iso-Citrate³⁻ + H⁺	α-Ketoglutarate²⁻ + CO_2 (gas)	2.32	0.363	Footnote by Harary, Korey and Ochoa (1953), Burton and Wilson (1953).
2 Cysteine	Cystine	3.42	−0.340	Kolthoff et al. (1955).
Cytochromes: 2 Fe^{2+} + 2 H⁺	3 Fe^{3+}			
a	(p_H 7.4)	33.5	0.29	Wainio (1955), Ball (1938).
b	(p_H 7.4)	18.3	−0.04	Ball (1938).
b_5		18.5	−0.12	Strittmatter and Ball (1952).
c		30.7	0.25	Paul (1947).
f		35.9	0.365	Davenport and Hill (1952).
DPNH + H⁺	DPN⁺	4.33	−0.320	Burton and Wilson (1953).
Ethanol	Acetaldehyde	10.01	−0.197	From Table 9.
FADH + H⁺	FAD	9.9	−0.20	Approximate. Dixon (1949).
Glucose + H_2O	Gluconate⁻ + H⁺	−2.7	−0.47	Strecker and Korkes (1952).
Glucose	Gluconolactone	2.30	−0.364	Burton and Wilson (1953).
Glucose	Gluconolactone	3.66	−0.335	Brink (1953), Burton and Wilson (1953).
Glutamate⁻ + H_2O	α-Ketoglutarate²⁻ + NH_4^+	12.94	−0.133	From Table 9.
Glutathione. 2GSH	G—S—S—G	3.42	−0.340	Kolthoff et al. (1955).
Glyceraldehyde 3-phosphate²⁻ + HPO_4^{2-}	Glyceroylphosphate 3-phosphate⁴⁻	5.83	−0.286	Burton and Wilson (1953).
Glycerate⁻	Hydroxypyruvate⁻	11.78	−0.158	Zelitch (1955).
Glycerol 1-phosphate²⁻	Dihydroxyacetone phosphate²⁻	10.24	−0.192	Burton and Wilson (1953).
Glycollate⁻	Glyoxylate⁻	14.88	−0.090	Zelitch (1955).
Glyoxylate⁻	Oxalate²⁻ + H⁺	−2.2	−0.462	Approximate. From Table 9.

Table 10. (Continued)

Reactants	Products (+ H$_2$ gas)	$\Delta G'$ (kgcal) at pH 7	E_0' (volts) at pH 7	References and comments
H$_2$O$_2$	O$_2$ (gas)	32.7	0.295	Approximate. From Table 9.
H$_2$S (gas)	S (rhombic)	7.89	−0.243	From Table 9.
HS$^-$ + H$^+$	S (rhombic)	6.55	−0.272	
L-β-Hydroxybutyrate$^-$	Acetoacetate$^-$	3.0	−0.349	Green, Dewan and Leloir (1937), Burton and Wilson (1953).
D-β-Hydroxybutyryl CoA	Acetoacetyl CoA	8.1	−0.238	Lynen and Wieland (1955), Burton and Wilson (1953).
Hypoxanthine + H$_2$O	Xanthine	2.0	−0.37	Green (1934).
Lactate$^-$	Pyruvate$^-$	10.32	−0.190	Burton and Wilson (1953).
Malate^{2-}	Oxaloacetate^{2-}	11.45	−0.166	Burton and Wilson (1953).
Malate^{2-} + H$^+$	Pyruvate$^-$ + CO$_2$ (gas)	3.83	−0.330	From Table 9.
NO$_2^-$ + H$_2$O	NO$_3^-$ + H$_2$	38.5	0.421	From Table 9.
α-Ketoglutarate^{2-} + H$_2$O	Succinate^{2-} + CO$_2$ (gas)	−11.92	−0.673	From Table 9.
iso-Propanol	Acetone	5.89	−0.296	Burton and Wilson (1953).
Pyruvate$^-$ + H$_2$O	Acetate$^-$ + CO$_2$ (gas)	−13.12	−0.699	From Table 9.
Riboflavin	Leucoriboflavin	9.9	−0.20	Kuhn and Boulanger (1936), Michaelis, Schubert and Smythe (1936).
Sorbitol	Fructose	6.53	−0.272	Blakley (1951), Burton and Wilson (1953).
Succinate^{2-}	Fumarate^{2-}	20.5	0.031	Burton and Krebs (1953).
SO$_3^{2-}$ + H$_2$O	SO$_4^{2-}$	−1.9	−0.454	Rossini et al. (1952).
½ S$_2$O$_4^{2-}$ + H$_2$O	SO$_3^{2-}$ + H$^+$	−2.7	−0.471	
S$_2$O$_3^{2-}$ + H$_2$O	S$_2$O$_4^{2-}$	41.4	0.484	
TPNH + H$^+$	TPN$^+$	4.11	−0.324	Burton and Wilson (1953), Kaplan, Colowick and Neufeld (1953), Black and Wright (1955).
Water (H$_2$O)	½ O$_2$	56.69	0.816	Rossini et al. (1952).
Xanthine + H$_2$O	Urate$^-$ + H$^+$	2.5	−0.36	Green (1934).
Yellow enzyme. Reduced	Oxidized	13.5	−0.122	Vestling (1955).

The free energies of formation (Table 9) have, where necessary, been recalculated so that the values for the basic ancillary data (atomic weights, entropies of the elements and heats of formation of CO$_2$ and H$_2$O) are the same as those used by Rossini, Wagman, Evans, Levine and Jaffe (1952). No Washburn corrections have been applied to combustion data which have not already been thus corrected, but such data are indicated by special comments. Similarly, corrections have not been introduced into the combustion data for the slight changes which have been made in the accepted heat of combustion of benzoic acid, which is the calorimetric standard. Any such corrections are negligible in the data given in these tables.

Table 11. *Glycolysis and alcoholic fermentation*
Values are for aqueous solution except that CO_2 and O_2 are gases at 1 atmosphere

	$\Delta G'$ (kcal) at PH 7	Reference
Glycogen (1 C_6 unit) + $H_2O \to$ 2 lactate$^-$ + 2 H$^+$	−51.6	From Table 9.
Glucose $\to$ 2 lactate$^-$ + 2 H$^+$	−47.4	From Table 9.
Glucose $\to$ 2 ethanol + 2 CO_2	−56.1	From Table 9.
Glucose + ATP$^{4-} \to$ glucose 6-P^{2-} + ADP^{3-} + H$^+$	−5.1	Burton and Krebs (1953), Burton (1955).
Glucose 6-P$^{2-} \to$ glucose 1-P^{2-}	+1.72	Colowick and Sutherland (1942).
Glucose 1-P^{2-} + $H_2O \to$ glycogen + HPO_4^{2-}	−0.55	Burton and Krebs (1953).
Glucose 6-P$^{2-} \to$ fructose 6-P^{2-}	+0.50	Slein (1950).
Fructose 6-P^{2-} + ATP$^{4-} \to$ fructose 1:6-diphosphate^{4-} + ADP^{3-} + H$^+$	−4.2	Burton and Krebs (1953), Burton (1955).
Fructose 1:6-diphosphate$^{4-} \to$ glyceraldehyde 3-P^{2-} + dihydroxyacetone P^{2-}	+5.51	Meyerhof and Lohmann (1934).
Dihydroxyacetone P$^{2-} \to$ glyceraldehyde 3-P^{2-}	+1.83	Meyerhof and Junowicz-Kocholaty (1943).
Glyceraldehyde 3 P^{2-} + DPN$^+$ + $HPO_4^{2-} \to$ glyceroyl-P 3-P^{4-} + DPNH + H$^+$	+1.50	Burton and Wilson (1953).
Glyceroyl-P 3-P^{4-} + ADP$^{3-} \to$ glycerate 3-P^{3-} + ATP^{4-}	−4.75	Bücher (1947).
Glycerate 3-P$^{3-} \to$ glycerate 2-P^{3-}	+1.06	Meyerhof and Oesper (1947).
Glycerate 2-P$^{3-} \to$ enolpyruvate 2-P^{3-} + H_2O	−0.64	Meyerhof and Oesper (1947).
Enolpyruvate 2-P^{3-} + ADP^{3-} + H$^+ \to$ pyruvate + ATP^{4-}	−6.1	Burton (1955).
Pyruvate$^-$ + DPNH + H$^+ \to$ lactate$^-$ + DPN$^+$	−6.0	Racker (1950).
Pyruvate$^-$ + H$^+ \to$ acetylaldehyde + CO_2	−5.1	From Table 9.
Acetylaldehyde + DPNH + H$^+ \to$ ethanol + DPN$^+$	−5.4	Racker (1950).

Table 12. *Tricarboxylic acid cycle*

Values are for aqueous solutions except that CO_2 and O_2 are gases at 1 atmosphere

	$\Delta G'$ (kcal) (pH 7)	Reference
Pyruvate$^-$ + ½O_2 + CoA + H$^+$ → acetyl CoA + H_2O + CO_2	− 61.8	Table 9 and Burton (1955).
Oxaloacetate^{2-} + acetyl CoA + H_2O → citrate^{3-} + CoA + H$^+$	− 7.5	Stern, Ochoa and Lynen (1952).
Citrate^3 → cis-aconitate^{3-} + H_2O	+ 2.04	Krebs (1953). (a)
cis-Aconitate^{3-} + H_2O → isocitrate^{3-}	− 0.45	Krebs (1953). (a)
iso-Citrate^{3-} + ½O_2 + H$^+$ → α-ketoglutarate^{2-} + H_2O + CO_2	− 54.4	Footnote by Harary et al. (1953), Burton and Wilson (1953).
α-Ketoglutarate^{2-} + ½O_2 + CoA + H$^+$ → succinyl CoA$^-$ + H_2O + CO_2	− 59.6	Kaufman and Alivisatos (1955), Burton (1955) and Table 9.
Succinyl CoA$^-$ + ADP^{3-} + HPO$_4^{2-}$ → succinate^{2-} + ATP^{4-} + CoA	− 0.77	Kaufman and Alivisatos (1955), Burton (1955) and Table 9.
Succinate^{2-} + ½O_2 → fumarate^{2-} + H_2O	− 36.1	Table 9.
Fumarate^{2-} + H_2O → malate^{2-}	− 0.88	Krebs [1953(a)].
Malate^{2-} + ½O_2 → oxaloacetate^{2-} + H_2O	− 45.3	Burton and Wilson (1953) and Table 9.
Malate^{2-} + TPN$^+$ → pyruvate$^-$ + TPNH + CO_2	+ 0.3	Burton and Wilson (1953) and Table 9.
DPNH + ½O_2 + H$^+$ → DPN$^+$ + H_2O	− 52.4	Burton and Wilson (1953) and Table 9.
TPNH + ½O_2 + H$^+$ → TPNH$^+$ + H_2O	− 52.6	Burton and Wilson (1953) and Table 9.
Glucose + 6 O_2 → 6 CO_2 + 6 H_2O	− 686.5	Table 9.
Pyruvate$^-$ + 2½O_2 + H$^+$ → 3 CO_2 + 2 H_2O	− 273.2	Table 9.

Table 13. *Free energies of hydrolysis*

The values for the third to the seventh reactions are based on those for the hydrolysis of ATP, and/or acetyl CoA

Reactant (+ H₂O)	Products	p_H	$\Delta G'$ (kgcal)	Reference
ATP⁴⁻ →	ADP³⁻ + HPO₄²⁻ + H⁺	7.5	—8.9	Burton (1955).
Acetyl CoA →	acetate⁻ + CoA + H⁺	7.0	—8.2	Burton (1955).
ADP³⁻ →	AMP²⁻ + HPO₄²⁻ + H⁺	7.5	—9.5	Burton and Krebs (1953).
Propionyl CoA →	propionate⁻ + CoA + H⁺	7.0	—8.2	Stadtman and Barker (1950).
Heptanoyl CoA →	heptanoate⁻ + CoA + H⁺	7.0	—8.8	Mahler, Wakil and Bock (1953).
Succinyl CoA⁻ →	succinate²⁻ + CoA + H⁺	7.0	—9.0	Kaufman and Alivisatos (1955).
HP₂O₇³⁻ →	2 HPO₄²⁻ + H⁺	7.5	—8.9	Jones (1953).
Alanyl glycine →	alanine + glycine		—4.13	Borsook (1954). $\Delta G'$ is p_H independent.
Glycyl glycine →	2 glycine		—3.59	Borsook (1954). $\Delta G'$ is p_H independent.
Leucyl glycine →	leucine + glycine		—3.31	Borsook (1954). $\Delta G'$ is p_H independent.
Hippurate⁻ →	benzoate⁻ + glycine		—2.63	Borsook (1954). $\Delta G'$ is p_H independent.
N-benzoyl tyrosine glycineamide →	N-benzoyl-tyrosine + glycineamide		—0.36	Borsook (1954). $\Delta G'$ is p_H independent.

The free energies and potentials of oxidation-reduction reactions (Table 10) enable the free energy changes to be readily calculated for many reactions which involve a transfer of hydrogen or of electrons. The free energy changes in Table 10 are those for the oxidation of the reactants in the first column and their conversion to the products in the second column with the concomitant formation of one molecule of gaseous H_2 (or the thermodynamically equivalent $2 H^+ + 2$ electrons, at unit activity). The values for any two of the reactions may, of course, be used to obtain $\Delta G'$ for a reaction which may be represented as their sum. Thus, if $\Delta G'$ (+32.7 kgcal) for the reaction

$$H_2O_2 = O_2 + H_2$$

is subtracted from $\Delta G'$ (+13.0 kgcal) for the reaction

$$\text{alanine} + H_2O = \text{pyruvate} + NH_4^+ + H_2$$

the value of $(13.0 - 32.7) = -19.7$ kgcal is obtained for $\Delta G'$ of the reaction

$$\text{alanine} + H_2O + O_2 = \text{pyruvate} + NH_4^+ + H_2O_2.$$

The relation between the oxidation-reduction potential (E_0') and $\Delta G'$ at the same p_H is

$$E_0' = \frac{\Delta G'}{2F} - \frac{2.3026 RT\, p_H}{F}$$

where F is the Faraday, R the gas constant and T the absolute temperature. If $\Delta G'$ is in kgcal and E_0' in volts.

$$E_0' = 0.0217\,\Delta G' - 0.0591\, p_H \text{ at } 25°.$$

Oxidation-reduction data are available for other compounds not listed here. Values for oxidation-reduction indicators and for many biological reactions are given by CLARK (1948) and by JOHNSON (1949); data for inorganic reactions by LATIMER (1952) and ROSSINI et al. (1952).

Tables 11 and 12 list the free energy changes for the reactions of glycolysis and alcoholic fermentation, and for the reactions of the tricarboxylic acid cycle. Insufficient data are available (February 1956) to obtain consistent free energy values for the reactions of fatty acid oxidation; for various reasons it is possible that the value given in Table 10 for the dehydrogenation of butyryl CoA may be inaccurate and closer to that for the dehydrogenation of succinate. The estimates given in Table 13 for the free energies of hydrolysis of ATP, acetyl CoA and certain other "energy-rich" compounds are thought to be correct to $\pm$ 1.5 kgcal. The error is in part due to uncertainties in the activity coefficients of the components of the various equilibrium systems studied and in part to complex formation with magnesium ions in the equilibrium mixtures.

References

ALBERT, A., and D. J. BROWN: Purine studies. I. Stability to acid and alkali. Solubility, ionisation, comparison with pteridines. J. Chem. Soc. (Lond.) **1954**, 2060.

ARNON, D. I.: The chloroplast as a complete photosynthesis unit. Science (Lancaster, Pa.) **122**, 9 (1955).

— L. L. ROSENBERG and F. R. WHATLEY: A new glyceraldehyde phosphate dehydrogenase from photosynthetic tissues. Nature (Lond.) **173**, 1132 (1954).

AUBEL, E.: Remarques sur la croissance du bacille coli en milieu chimiquement défini. Ann. Physiol. et Physiochim. biol. **2**, 73 (1926).

AXELROD, B., and R. JANG: Purification and properties of phosphoriboisomerase from alfalfa. J. of Biol. Chem. **202**, 619 (1954).

BACH, S. J.: The metabolism of protein constituents in the mammalian body. Oxford: Clarendon Press 1952.

BALL, E. G.: Über die Oxydation und Reduktion der drei Cytochromokomponenten. Biochem. Z. **295**, 262 (1938).

— Energy relationships of the oxidative enzymes. Ann. New York Acad. Sci. **45**, 363 (1944).

BARANOWSKI, T.: Crystalline glycerophosphate dehydrogenase from rabbit muscle. J. of Biol. Chem. **180**, 535 (1949).

BEATTY, C. H., and E. S. WEST: The effect of substances related to the tricarboxylic acid cycle upon ketosis. J. of Biol. Chem. **190**, 603 (1951).

BEINERT, H., D. E. GREEN, P. HELE, H. HIFT, R. W. v. KORFF and C. V. RAMAKRISHNAN: The acetate activating enzyme system of heart muscle. J. of Biol. Chem. **203**, 35 (1953).

BELITZER, V. A.: La régulation de la respiration musculaire par les transformations du phosphogène. Enzymologia (Den Haag) **6**, 1 (1939).

BERG, P.: Participation of adenyl-acetate in the acetate-activating system. J. Amer. Chem. Soc. **77**, 3163 (1955).

—, and W. K. JOKLIK: Transphorylation between nucleoside polyphosphates. Nature (Lond.) **172**, 1008 (1953).

— — Enzymic phosphorylation of nucleoside diphosphate. J. of Biol. Chem. **210**, 657 (1954).

BERNAL, J. D.: The origin of life. New Biology, vol. 16, p. 28. London: Penguin Books 1954.
BLACK, S., and N. G. WRIGHT: Homoserine dehydrogenase. J. of Biol. Chem. **213**, 51 (1955).
BLAKLEY, R. L.: The metabolism and antiketogenic effects of sorbitol. Sorbitol dehydrogenase. Biochemic. J. **49**, 257 (1951).
BLUM, J. F.: Time's arrow and evolution. Princeton, New Jersey: Princeton University Press 1951.
BORGSTRÖM, B., H. C. SUDDRUTH and A. L. LEHNINGER: Phosphorylation coupled to the reduction of cytochrome c by β-hydroxybutyrate. J. of Biol. Chem. **215**, 571 (1955).
BORSOOK, H.: Enzymatic syntheses of peptide bonds. In: Chemical pathways of metabolism, vol. 2, p. 173, edit. by D. M. GREENBERG. New York: Academic Press Inc. 1954.
— H. W. DAVENPORT, C. E. P. JEFFREYS and R. C. WARNER: The oxidation of ascorbic acid and its reduction in vitro and in vivo. J. of Biol. Chem. **117**, 237 (1937).
— E. L. ELLIS and H. M. HUFFMAN: Sulphydryl oxidation reduction potentials derived from thermal data. J. of Biol. Chem. **117**, 281 (1937).
—, and H. M. HUFFMAN: Some thermodynamical considerations of amino acids, peptides and related substances. In: Chemistry of the amino acids and proteins, edit. by C. L. A. SCHMIDT. Springfield: Ch. C. Thomas 1938.
BOYER, P. D., and H. L. SEGAL: Sulphydryl groups of glyceraldehyde 3-phosphate dehydrogenase and acyl enzyme formation. In: A symposium on the mechanism of enzyme action, edit. by W. D. MCELROY and B. GLASS. Baltimore: Johns Hopkins Press 1954.
BRAUNSTEIN, A. E.: Transamination and the integrative functions of the dicarboxylic acids in nitrogen metabolism. Adv. Protein Chem. **3**, 1 (1947).
BREWER, C. R., and C. H. WERKMAN: The aerobic dissimilation of citric acid by coliform bacteria. Enzymologia (Den Haag) **8**, 318 (1940).
BRINK, N. G.: Beef liver glucose dehydrogenase. I. Purification and properties. Acta chem. scand. (Copenh.) **7**, 1081 (1953).
BRODIE, A., and F. LIPMANN: Identification of a gluconolactonase. J. of Biol. Chem. **212**, 677 (1955).
BÜCHER, T.: Über ein phosphatübertragendes Gärungsferment. Biochim. et Biophysica. Acta **1**, 292 (1947).
BULLOCK, M. W., J. A. BROCKMAN jr., E. L. PATTERSON, J. V. PIERCE and E. L. R. STOKSTAD: Sythesis of DL-thioctic acid, J. Amer. Chem. Soc. **74**, 1868, 3455 (1952).
BURK, D.: A colloquial consideration of the Pasteur and Neo-Pasteur effects. Cold Spring Harbor Symp. on Quant. Biol. **7**, 420 (1939).
BURTON, K.: The free-energy change associated with the hydrolysis of acetyl coenzyme A. Biochemic. J. **59**, 44 (1955).
—, and H. A. KREBS: The free-energy changes associated with the individual steps of the tricarboxylic acid cycle, glycolysis and alcoholic fermentation and with the hydrolysis of the pyrophosphate groups of adenosine-triphosphate. Biochemic. J. **54**, 94 (1953).
—, and T. H. WILSON: The free-energy changes for the reduction of diphosphopyridine nucleotide and the dehydrogenation of L-malate and L-glycerol 1-phosphate. Biochemic. J. **54**, 86 (1953).
BURTON, R. M., and E. R. STADTMAN: The oxidation of acetaldehyde to acetyl coenzyme A. J. of Biol. Chem. **202**, 873 (1953).
BUTLER, J. A. V., C. N. RAMCHANDANI and D. W. THOMSON: The solubility of non-electrolytes. Part I. The free-energy of hydration of some aliphatic alcohols. J. Chem. Soc. (Lond.) **1935**, 280.
CALVIN, M.: The photosynthetic carbon cycle. Proc. 3rd. Int. Congr. of Biochem., Brussels 1955, edit. by C. LIEBECQ, p. 211. New York: Academic Press 1956.
— J. R. QUAYLE, R. C. FULLER, J. MAYAUDON, A. A. BENSON and J. A. BASSHAM: The photosynthetic carbon cycle. Federat. Proc. **14**, 188 (1955).
CAMMARATA, P. S., and P. P. COHEN: The scope of the transamination reaction in animal tissues. J. of Biol. Chem. **187**, 439 (1950).

CAMPBELL, J. J. R., and I. C. GUNSALUS: Citric acid fermentation by streptococci and lactobacilli. J. Bacter. **48**, 71 (1944).
CHANCE, B.: Enzymes in action in living cells: the steady state of reduced pyridine nucleotides. Harvey Lect. **49**, 145 (1955).
CHOU, T. C., and F. LIPMANN: Separation of acetyl transfer enzymes in pigeon liver extract. J. of Biol. Chem. **196**, 89 (1952).
CLARK, W. M.: Topics in physical chemistry, 1st edit. London: Ballière, Tyndall and Cox 1948.
CLARKE, E. W., and B. C. WHALER: The utilization of ^{14}C-labelled amino acids by the isolated mammalian heart. J. of Physiol. **117**, 9P (1952).
CLARKE, T. H., and G. STEGEMAN: Heats of combustion of some mono- and disaccharides. J. Amer. Chem. Soc. **61**, 1726 (1939).
COHEN, G. N.: Nature et mode de formation des acides volatils trouvés dans les cultures de bactéries anaérobies strictes. Ann. Inst. Pasteur **77**, 471 (1949).
—, and G. COHEN-BAZIRE: Fermentation of pyruvate, β-hydroxybutyrate and of C_4-dicarboxylic acids by some butyric acid forming, organisms. Nature (Lond.) **162**, 578 (1948).
COHEN, S. S.: Comparative biochemistry and chemotherapy. In: Cellular metabolism and infections, p. 84, edit. by E. RACKER. New York: Academic Press, Inc. 1954.
COHEN-BAZIRE, G., and G. N. COHEN: Études sur le mécanisme de la fermentation acétono-butylique. I. Synthèse d'acide butyrique à partir de pyruvate. Ann. Inst. Pasteur **77**, 718 (1949).
— — B. NISMAN et M. RAYNAUD: Action inhibitrice d l'arsénite de sodium sur la production d'acide butyrique á partir de pyruvate, chez. Cl. sacharobutyricum. C. r. Soc. Biol. Paris **142**, 1221 (1948).
COHN, M.: Phosphorus metabolism, vol. 1,p. 374, edit. by W. D. McELROY and B. GLASS. Baltimore: Johns Hopkins Press 1951.
— Some mechanisms of clearage of adenosine triphosphate and 1,3 phosphoglyceric acid. Biochim. et Biophysica Acta **20**, 92 (1956).
COLOWICK, S. P., and E. W. SUTHERLAND: Polysaccharide synthesis from glucose by means of purified enzymes. J. of Biol. Chem. **144**, 423 (1942).
COOK, R. P.: Pyruvic acid in bacterial metabolism. Biochemic. J. **24**, 1526 (1930).
COON, M. J., W. G. ROBINSON and B. K. BACHHAWAT: Enzymatic studies on the biological degradation of the branched chain amino acids. In: Symposium on amino acid metabolism, p. 431, edit. by W. D. McELROY and H. B. GLASS. Baltimore: Johns Hopkins Press 1955.
COOPER, C., T. M. DEVLIN and A. L. LEHNINGER: Oxidative phosphorylation in an enzyme fraction from mitochondrial extracts. Biochim. et Biophysica Acta **18**, 159 (1955).
CORI, G. T., and C. F. CORI: Glucose 6-phosphatase of the liver in glycogen storage disease. J. of Biol. Chem. **199**, 661 (1952).
CORI, O., and F. LIPMANN: The primary oxidation product of enzymatic glucose 6-phosphate oxidation. J. of Biol. Chem. **194**, 417 (1952).
CRANDALL, D. I.: Homogentisic acid oxidase. J. of Biol. Chem. **212**, 565 (1955)(a).
— The ferrous ion activation of homogentisic acid oxidase and other aromatic ring-splitting oxidases. In: Symposium on amino acid metabolism, p. 867, edit. by W. D. McELROY and H. B. GLASS. Baltimore: Johns Hopkins Press 1955.(b)
CRANE, R. K., and A. SOLS: The non-competitive inhibition of brain hexokinase by glucose 6-phosphate and related compounds. J. of Biol. Chem. **210**, 597 (1954).
DAUVILLIER, A., and E. DESGUIN: La genèse de la vie, phase de l'évolution géochimique. Paris: Hermans 1942.
DAVENPORT, H. E., and R. HILL: The preparation and some properties of cytochrome f. Proc. Roy. Soc. Lond., Ser. B **139**, 327 (1952).
DAVIES, D. D.: In press. Biochemic. J. **1956**.

Davies, R. E.: Relations between active transport and metabolism in some isolated tissues and mitochondria. In: Active transport and secretion. Sympos. Soc. Exper. Biol. 8, 453 (1954).
—, and H. A. Krebs: Biochemical aspects of the transport of ions by nervous tissue. Biochem. Soc. Symposia 1952, No 8, 77.
Davis, B. D.: Biosynthesis of the aromatic amino acids. In: A symposium on amino acid metabolism, p. 799, edit. by W. D. McElroy and B. Glass. Baltimore: Johns Hopkins Press. 1955
De la Haba, G. L., E. Racker and I. G. Leder: Crystalline transketolase from bakers' yeast: Isolation and properties. J. of Biol. Chem. 214, 409 (1955).
Deuel, H. R., S. Murray and L. F. Hallman: A comparison of the ketolytic effect of succinic acid with glucose. Proc. Soc. Exper. Biol. a. Med. 37, 413 (1937).
Dickens, F.: Mechanism. of carbohydrate oxydation. Nature (Lond.) 138, 1057 (1936).
— Oxidation of phosphohexonate and pentose phosphoric acids by yeast enzymes. I. Oxidation of phosphohexonate. II. Oxidation of pentose phosphoric acids. Biochemic. J. 32, 1626 (1938) (a).
— Yeast fermentation of pentose phosphoric acids. Biochemic. J. 32, 1645 (1938) (b).
—, and D. H. Williamson: Transformation of pentose phosphates by enzymes of animal origin. Nature (Lond.) 176, 400 (1955).
Dische, Z.: Phosphorylierung der im Adenosin enthaltenen d-Ribose nnd nachfolgender Zerfall des Esters unter Triosephosphatbildung im Blute. Naturwiss. 26, 252 (1938).
— Synthesis of hexosemono- and diphosphate from adenosine and ribose-5-phosphate in human blood. In: Phosphorus metabolism, vol. I, p. 171, edit. by W. D. McElroy and B. Glass. Baltimore: Johns Hopkins Press 1951.
Dixon, M.: Multi-enzyme systems. Cambridge 1949.
Ehrensvärd, G.: Metabolism of amino acids and proteins. Annual Rev. Biochem. 24, 275 (1955).
Euler, H. v., E. Adler and G. Günther: Zur Kenntnis der Dehydrierung von α-Glycerin-phosphorsäure im Tierkörper. Z. physiol. Chem. 249, 1 (1937).
Fantl, P., and N. Rome: Dephosphorylation in liver extracts. Austral. J. Exper. Biol. a. Med. Sci. 23, 21 (1945).
Findlay, A.: The solubility of mannitol, picric acid and anthracene. J. Chem. Soc. (Lond.) 1902, 1217.
Gibbs, M., and B. L. Horecker: The mechanism of pentose phosphate conversion to hexose monophosphate. II. With pea leaf and pea root preparations. J. of Biol. Chem. 208, 813 (1954).
Gillespie, R. J., G. A. Maw and C. A. Vernon: The concept of phosphate bond energy. Nature (Lond.) 171, 1147 (1953).
Gilvarg, C.: Prephenic acid and the aromatization step in the synthesis of phenylalanine. In: A symposium on amino acid metabolism, p. 812, edit. by W. D. McElroy and B. Glass. Baltimore: Johns Hopkins Press 1955.
Glock, G. E., and P. McLean: The determination of oxidised and reduced diphosphopyridine nucleotide and triphosphopyridine nucleotide in animal tissues. Biochemic. J. 61, 381 (1955) (a).
— — Levels of oxidized and reduced diphosphopyridine nucleotide and triphosphopyridine nucleotide in animal tissues. Biochem. J. 61, 388 (1955) (b).
Goldschmidt, V. M.: Geochemical aspects of the origin of complex organic molecules on the earth, as precursors to organic life. New Biology, vol. 12, p. 97. London: Penguin Books 1952.
Gomori, G.: Hexosediphosphatase. J. of Biol. Chem. 148, 139 (1943).
Green, D. E.: Studies on reversible dehydrogenase systems. II. The reversibility of the xanthine oxidase system. Biochemic. J. 28, 1550 (1934).
— Enzymes in metabolic sequences. In: Chemical pathways of metabolism, vol. 1, p. 27, edit. by D. M. Greenberg. 1954 (a).
— Fatty acid oxidation in soluble systems of animal tissues. Biol. Rev. 29, 330 (1954) (b).

GREEN, D. E.: Organization in relation to enzymatic functions. Symposium of the society for experimental biology on mitochondria and other protoplasmic inclusions. In press (1955).
— J. G. DEWAN and L. F. LELOIR: The β-hydroxybutyric dehydrogenase of animal tissues. Biochemic. J. **31**, 934 (1937).
— S. MII and P. M. KOHOUT: Studies on the terminal electron transport system. I. Succinic dehydrogenase. J. of Biol. Chem **217**, 551 (1955).
— — H. R. MAHLER and R. M. BOCK: Studies on the fatty acid oxidising system of animal tissues. III. Butyryl coenzyme A dehydrogenase. J. of Biol. Chem. **206**, 1 (1954).
GREENBERG, D. M.: Carbon catabolism of amino acids. In: Chemical pathways of metabolism, vol. II, p. 47. 1954.
GUNSALUS, I. C.: Products of aerobic glycerol fermentation by streptococcus faecalis. J. Bacter. **54**, 239 (1947).
— Group transfer and acyl-generating functions of lipoic acid derivatives. In: The mechanism of enzyme action, p. 545, edit. by W. D. MCELROY and B. GLASS. Baltimore: Johns Hopkins Press 1954 (a).
— Oxidative and transfer reactions of lipoic acid. Federat. Proc. **13**, 715 (1954) (b).
— B. L. HORECKER and W. A. WOOD: Pathways of carbohydrate metabolism in microorganisms. Bacter. Rev. **19**, 79 (1955).
HAGER, L. P., and I. C. GUNSALUS: Lipoic acid dehydrogenase: The function of E. coli fraction B. J. Amer. Chem. Soc. **75**, 5767 (1953).
HALDANE, J. B. S.: The origin of life. Rationalist Annual. 1929. Reprinted in: The inequality of man. London 1932. Published as Science and human life. New York 1933.
— The origins of life. New Biology, vol. 16, p. 12. London: Penguin Books 1954.
HALDANE, J. S., and J. G. PRIESTLEY: The regulation of the lung ventilation. J. of Physiol. **32**, 225 (1905).
HANSEN, R. S., F. A. MILLER and S. D. CHRISTIAN: Activity coefficients of components to the systems water-acetic acid, water-propionic acid and water-n-butyric acid at 25°. J. of Physic. Chem. **59**, 391 (1955).
HARARY, I., S. R. KOREY and S. OCHOA: Biosynthesis of dicarboxylic acids by carbon dioxide fixation. VII. Equilibrium of "malic" enzyme reaction. J. of Biol. Chem. **203**, 595 (1953).
HARDEN, A.: The chemical action of "B. coli communis" and similar organisms on carbohydrates and allied compounds. Trans. Chem. Soc. **79**, 610 (1901).
HERBERT, D.: Oxidising enzymes. Ann. Rep. Progr. Chem. **47**, 335 (1951).
HERSEY, D. F., and S. J. AJL: Adenosine triphosphate formation in the oxidation of succinic acid by bacteria. J. Gen. Physiol. **34**, 295 (1951).
HILLS, G. M.: Ammonia production by pathogenic bacteria. Biochemic. J. **34**, 1057 (1940).
HOLZER, H.: Über Fermentketten und ihre Bedeutung für die Regulation des Kohlenhydratstoffwechsels in lebenden Zellen. 4. Kolloquium der Ges. für physiologische Chemie, S. 89. 1953.
— Kinetik und Thermodynamik enzymatischer Reaktionen in lebenden Zellen und Geweben. In: Ergebnisse der Medizinischen Grundlagenforschung, S. 191. Stuttgart: Georg Thieme 1956.
HORECKER, B. L., M. GIBBS, H. KLENOW and P. Z. SMYRNIOTIS: The mechanism of pentose phosphate conversion to hexose monophosphate. I. With a liver enzyme preparation. J. of Biol. Chem. **207**, 393 (1954).
— J. HURWITZ and P. Z. SMYRNIOTIS: Xylulose 5-phosphate and the formation of sedoheptulose 7-phosphate with liver transketolase. J. Amer. Chem. Soc. **78**, 692 (1956).
— — and A. WEISSBACH: The enzymatic synthesis and properties of ribulose 1,5 diphosphate. J. of Biol. Chem. **218**, 785 (1956).

HORECKER, B. L., and P. Z. SMYRNIOTIS: Transaldolase: The formation of fructose 6-phosphate from sedoheptulose 7-phosphate. J. Amer. Chem. Soc. **75**, 2021 (1953).
— — Purification and properties of yeast transaldolase. J. of Biol. Chem. **212**, 811 (1952).
— — H. HIATT and P. MARKS: Tetrose phosphate and the formation of sedoheptulose diphosphate. J. of Biol. Chem. **212**, 827 (1955).
— — and J. E. SEEGMILLER: The enzymatic conversion of 6-phosphogluconate to ribulose 5-phosphate and ribose 5-phosphate. J. of Biol. Chem. **193**, 383 (1951).
HOROWITZ, N. W.: On the evolution of biochemical syntheses. Proc. Nat. Acad. Sci. U.S.A. **31**, 153 (1945).
HUDSON, C. S., and F. C. BROWN: The heats of solution of the three forms of milk-sugar. J. Amer. Chem. Soc. **30**, 960 (1908).
HUFFMAN, H. M.: Thermal data XV. The heats of combustion and free energies of some compounds containing the peptide bond. J. of Physic. Chem. **46**, 885 (1942).
—, and E. L. ELLIS: Thermal data VIII. The heat capacities, entropies and free energies of some amino acids. J. Amer. Chem. Soc. **59**, 2150 (1937).
—, and S. W. FOX: Thermal data X. Heats of combustion and free energies of some organic compounds concerned in carbohydrate metabolism. J. Amer. Chem. Soc. **60**, 1400 (1938).
HURWITZ, J.: Conversion of ribulose 5-phosphate to ribulose 1:5 diphosphate. Federat. Proc. **14**, 230 (1955).
— A. WEISSBACH, B. L. HORECKER and P. Z. SMYRNIOTIS: Spinach phosphoribulokinase. J. of Biol. Chem. **218**, 769 (1956).
HYNDMAN, L. A., R. H. BURRIS and P. W. WILSON: Properties of hydrogenase from azotobacter vinelandii. J. Bacter. **65**, 522 (1953).
JACK, G. W., and G. STEGEMAN: The heat capacities and entropies of two monosaccharides. J. Amer. Chem. Soc. **63**, 2121 (1941).
JAKOBY, W. B., D. O. BRUMMOND and S. OCHOA: Formation of 3-phosphoglyceric acid by carbon dioxide fixation with spinach leaf enzymes. J. of Biol. Chem. **218**, 811 (1956).
JOHNSON, M. J.: The role of aerobic phosphorylation in the Pasteur effect. Science (Lancaster, Pa.) **94**, 200 (1941).
— In LARDY, H. A., Respiratory Enzymes, 2nd edit. Minneapolis: Burgess Publishing Co. 1949.
JONES, M. E.: Discussion in symposium on chemistry and functions of coenzyme A. Federat. Proc. **12**, 708 (1953).
— L. SPECTOR and F. LIPMANN: Carbamyl phosphate, the carbamyl donor in enzymatic citrulline synthesis. J. Amer. Chem. Soc. **77**, 819 (1955)(a).
— — — Carbamyl phosphate. Proc. 3rd. Int. Congr. of Biochem., Brussels 1955, edit. by V. LIÉBECQ, p. 278. New York: Academic Press 1956.
KALCKAR, H. M.: In: A symposium on the mechanism of enzyme action, p. 739, edit. by W. D. MCELROY and B. GLASS. Baltimore: Johns Hopkins Press 1954.
KAPLAN, N. O.: Thermodynamics and mechanism of the phosphate bond. In J. B. SUMNER and K. MYRBÄCK, The enzymes: Chemistry and mechanism of action, vol. 2, part. 1, p. 55. New York: Academic Press 1951.
— S. P. COLOWICK and E. F. NEUFELD: Pyridine nucleotide transhydrogenase. II. Direct evidence for and mechanism of the transhydrogenase reaction. J. of Biol. Chem. **195**, 107 (1952).
— — — Pyridine nucleotide transhydrogenase III. Animal tissue transhydrogenases. J. of Biol. Chem. **205**, 1 (1953).
KAUFMAN, S.: Studies on the mechanism of the reaction catalysed by the phosphorylating enzyme. J. of Biol. Chem. **216**, 153 (1955).
—, and S. G. A. ALIVISATOS: Purification and properties of the phosphorylating enzyme from spinach. J. of Biol. Chem. **216**, 141 (1955).
KEARNEY, E. B., and T. P. SINGER: On the prosthetic group of succinic dehydrogenase. Biochim. et Biophysica Acta **17**, 596 (1955).

KEMPNER, W., and F. KUBOWITZ: Wirkung des Lichtes auf die Kohlenoxydhemmung der Buttersäuregärung. Biochem. Z. **265**, 245 (1933).
KLUYVER, A. J.: The chemical activities of microorganisms. London: University of London Press 1931.
KNIVETT, V. A.: Citrulline as an intermediate in the breakdown of arginine by streptococcus faecalis. Biochemic. J. **50**, XXX (1952).
KNOX, W. E.: The metabolism of phenylalanine and tyrosine. In: Symposium on amino acid metabolism, p. 836, edit. by W. D. MCELROY and B. GLASS. Baltimore: Johns Hopkins Press 1955.
—, and S. W. EDWARDS: Homogentisate oxidase of liver. J. of Biol. Chem. **216**, 479 (1955) (a).
— — The properties of maleyl acetoacetate, the initial product of homogentisate oxidation in liver. J. of Biol. Chem. **216**, 459 (1955) (b).
KOLTHOFF, I. M., W. STRICKS and T. TANAKA: The polarographic prewaves of cystine (RSSR) and dithiodiglycollic acid (TSST) and the oxidation potentials of the systems RSSR—RSH and TSST—TSH. J. Amer. Chem. Soc. **77**, 4739 (1955).
KORÁNYI, A., and A. SZENT GYÖRGYI: Über die Bernsteinsäurebehandlung diabetischer Azidose. Dtsch. med. Wschr. **1937**, 1029.
KORNBERG, A., and W. E. PRICER jr.: Enzymic synthesis of the coenzyme A derivatives of long chain fatty acids. J. of Biol. Chem. **204**, 329 (1953) (a).
— — Enzymic esterification of α-glycerophosphate by long chain fatty acids. J. of Biol. Chem. **204**, 345 (1953) (b).
KORNBERG, H. L., and E. RACKER: Enzymic reactions of erythrose 4-phosphate. Biochemic. J. **61**, iij (1955).
KREBS, H. A.: The role of fumarate in the respiration of Bacterium coli commune. J. of Biochem. **31**, 2095 (1937).
— The intermediary stages in the biological oxidation of carbohydrate. Adv. Enzymol. **3**, 191 (1943).
— Cyclic processes in living matter. Enzymologia (Den Haag) **12**, 88 (1947).
— The tricarboxylic acid cycle. Harvey Lect. **44**, 165 (1950).
— Oxidation of amino acids. In J. B. SUMNER and K. MYRBÄCK, The Enzymes: Chemistry and mechanism of action, vol. 2, part 1, p. 499. New York: Academic Press 1951.
— The equilibrium constants of the fumarase and aconitase systems. Biochemic. J. **54**, 78 (1953) (a).
— Some aspects of the energy transformation in living matter. Brit. Med. Bull. **9**, 97 (1953) (b).
— The tricarboxylic acid cycle. In D. M. GREENBERG, Chemical pathways of metabolism, vol. I, p. 109. New York: Academic Press 1954 (a).
— Energy production in animal tissues and in micro-organisms. In: Cellular Metabolism and Infections by E. RACKER. New York: Academic Press, Inc. 1954 (b).
— Considerations concerning the pathways of syntheses in living matter. Synthesis of glycogen from non-carbohydrate precursors. Bull. Johns Hopkins Hosp. **95**, 19 (1954) (c).
— L. V. EGGLESTON and V. A. KNIVETT: Arsenolysis and phosphorolysis of citrulline in mammalian liver. Biochemic. J. **59**, 185 (1955).
—, and R. HEMS: Some reactions of adenosine and inosine phosphates in animal tissues. Biochim. et Biophysica Acta **12**, 172 (1953).
KUBOWITZ, F.: Über die Hemmung der Buttersäuregärung durch Kohlenoxyd. Biochem. Z. **274**, 285 (1934).
KUHN, R., and P. BOULANGER: Beziehungen zwischen Reduktions-Oxydations-Potential und chemischer Konstitution der Flavine. Ber. dtsch. chem. Ges. B **69**, 1557 (1936).
KURY, J. W., A. J. ZIELEN and W. L. LATIMER: Heats of formation and entropies of HS^- and S^{--}. Potential of sulfide-sulfur couple. U. S. Atomic Energy Comm. UCRL-2108, 3 (1953). Zit. in Chem. Abstr. **48**, 35 (1954).

LANG, K.: Der intermediäre Stoffwechsel. Heidelberg: Springer 1952.
LANGDON, R. G.: The requirement of triphosphopyridine nucleotide in fatty acid synthesis. J. Amer. Chem. Soc. **77**, 5190 (1955).
LARDY, H. A.: The role of phosphate and metabolic control mechanisms. In The Biology of Phosphorus. State College Press, Michigan 1952.
—, and H. WELLMAN: Oxidative phosphorylations: role of inorganic phosphate and acceptor systems in control of metabolic rates. J. of Biol. Chem. **195**, 215 (1952).
LATIMER, W. M.: Oxidation-potentials, 2nd edit. Prentice-Hall 1952.
LAWRENCE, R. D.: Diabetic ketosis and succinic acid. Lancet **1937 II**, 286.
— R. A. MCCANCE and N. ARCHER: Succinic acid treatment of diabetic ketosis. Brit. Med. J. **1937**, 214.
LEHNINGER, A. L.: A quantitative study of the products of fatty acid oxidation in liver suspensions. J. of Biol. Chem. **164**, 291 (1946).
— Esterification of inorganic phosphate coupled to electron transport between dihydrophosphopyridine nucleotide and oxygen. J. of Biol. Chem. **178**, 625 (1949).
— Oxidative phosphorylation in diphosphopyridine nucleotide-linked systems. In: Phosphorus Metabolism, vol. I, p. 344, edit. W. D. MCELROY and B. GLASS. Baltimore: Johns Hopkins Press 1951.
— Oxidative phosphorylation. Harvey Lect. **49**, 176 (1955).
LENNERSTRAND, A.: Über die Wirkung von Phosphat auf Oxydation und Phosphorylierung in durch Fluorid vergifteten Apo-Zymasesystem. Biochem. Z. **289**, 104 (1936).
LEPAGE, G. A.: A comparison of tumour and normal tissues with respect to factors affecting the rate of anaerobic glycolysis. Cancer Res. **10**, 77 (1950).
LERNER, A. B.: On the metabolism of phenylalanine and tyrosine. J. of Biol. Chem. **181**, 281 (1949).
LEWIS, G. N., and M. RANDALL: Thermodynamics, 1st edit. New York: McGraw-Hill Book Company, Inc. 1923.
LIPMANN, F.: Über die oxydative Hemmbarkeit der Glykolyse und den Mechanismus der PASTEURschen Reaktion. Biochem. Z. **265**, 133 (1933).
— Über die Hemmung der Mazerationssaftgärung durch Sauerstoff in Gegenwart positiver Oxydoreduktionssysteme. Biochem. Z. **268**, 205 (1934).
— Fermentation of phosphogluconic acid. Nature (Lond.) **138**, 588 (1936).
— Metabolic generation and utilization of phosphate bond energy. Adv. Enzymol. **1**, 99 (1941).
— Metabolic process patterns. In: Currents in biochemical research, p. 137, edit. by D. E. GREEN. New York: Interscience Publishers 1946.
LORBER, V., N. LIFSON, H. G. WOOD, W. SAKAMI and W. W. SHREEVE: Conversion of lactate to liver glycogen in the intact rat, studied with isotopic lactate. J. of Biol. Chem. **183**, 517 (1950).
— — W. SAKAMI and H. G. WOOD: Conversion of propionate to liver glycogen in the intact rat, studied with isotopic propionate. J. of Biol. Chem. **183**, 531 (1950).
LYNEN, F.: Über den aeroben Phosphatbedarf der Hefe. Ein Beitrag zur Kenntnis der PASTEURschen Reaktion. Liebigs Ann. **546**, 120 (1941).
— Acetyl coenzyme A and the fatty acid cycle. Harvey Lect. **48**, 210 (1954).
—, and R. KOENIGSBERGER: Zum Mechanismus der PASTEURschen Reaktion. Der Phosphat-Kreislauf in der Hefe und seine Beeinflussung durch 2,4-Dinitrophenol. Liebigs Ann. **573**, 60 (1951).
—, and O. WIELAND: β-Ketoreductase. In: Methods in Enzymology, vol. 1, p. 566. New York: Academic Press 1955.
MADISON, K. M.: The organism and its origin. Evolution **7**, 211 (1953).
MAHLER, H. R.: Role of coenzyme A in fatty acid metabolism. Federat. Proc. **12**, 694 (1953).
— S. J. WAKIL and R. M. BOCK: Studies on fatty acid oxidation. I. Enzymatic activation of fatty acids. J. of Biol. Chem. **204**, 453 (1953).

Martius, C.: Die Stellung des Phyllochinones (Vitamin K_1) in der Atmungskette. Biochem. Z. **326**, 26 (1954).
—, and D. Nitz-Litzow: Zum Wirkungsmechanismus des Vitamin K. Biochem. Z. **327**, 1 (1955).
McElroy, W. D.: Properties of the reaction utilizing adenosine triphosphate for bioluminescence. J. of Biol. Chem. **191**, 547 (1951).
—, and B. L. Strehler: Bioluminescence. Bacter. Rev. **18**, 177 (1954).
Mel, H. C.: Chemical thermodynamics of aqueous thiosulfate and bromate ions. U.S. Atomic Energy Comm. UCRL-2330, 2 (1953). Zit. in Chem. Abstr. **48**, 6228 (1954).
Metzler, D. E., J. B. Longenecker and E. E. Snell: The reversible catalytic cleavage of hydroxyaminoacids by pyridoxal and metal salts. J. Amer. Chem. Soc. **76**, 639 (1954).
— J. Olivard and E. E. Snell: Transamination of pyridoxamine and amino acids with glyoxylic acid. J. Amer. Chem. Soc. **76**, 644 (1954).
Meyerhof, O.: II. Das Schicksal der Milchsäure in der Erholungsperiode des Muskels. Pflügers Arch. **182**, 284 (1920).
—, and S. Fiala: Pasteur effect in dead yeast. Biochim. et Biophysica Acta **6**, 1 (1950).
—, and R. Junowicz-Kocholaty: The equilibria of isomerase and aldolase and the problem of the phosphorylation of glyceraldehyde phosphate. J. of Biol. Chem. **149**, 71 (1943).
—, and K. Lohmann: Über die enzymatische Gleichgewichtsreaktion zwischen Hexosediphosphorsäure und Dioxyacetonphosphorsäure. Biochem. Z. **271**, 89 (1934).
—, and P. Oesper: The mechanism of the oxidative reaction in fermentation. J. of Biol. Chem. **170**, 1 (1947).
Michaelis, L., M. P. Schubert and C. V. Smythe: Potentiometric study of the flavins. J. of Biol. Chem. **116**, 587 (1936).
Mii, S., and D. E. Green: Studies on the fatty acid oxidizing system of animal tissues. VIII. Reconstruction of fatty acid oxidizing system with triphenyl tetrazolium as electron acceptor. Biochim. et Biophysica Acta **13**, 425 (1954).
Miller, S. L.: A production of amino acids under possible primitive earth conditions. Science (Lancaster, Pa.) **117**, 528 (1953).
— Production of some organic compounds under possible primitive earth conditions. J. Amer. Chem. Soc. **77**, 2351 (1955).
Morales, M. F., J. Botts, J. J. Blum and T. L. Hill: Elementary processes in muscle action: An examination of current concepts. Physiologic. Rev. **35**, 475 (1955).
Nachmansohn, D.: Metabolism and function of the nerve cell. Harvey Lect. **49**, 57 (1955).
— C. W. Coates, M. A. Rothenberg and M. V. Brown: On the energy source of the action potential in the electric organ of Electrophorus electricus. J. of Biol. Chem. **165**, 223 (1946).
— R. T. Cox, C. W. Coates and A. L. Machado: Action potential and enzyme activity in the electric organ of Electrophorus electricus. II. Phospho-creatine as energy source of the action potential. J. of Neurophysiol. **6**, 383 (1943).
Niel, C. B. van: The comparative biochemistry of photosynthesis. In: Photosynthesis in Plants, p. 437, edit. J. Franck and W. E. Loomis. Ames: Iowa State College Press 1949.
Nielsen, S. O., and A. L. Lehninger: Phosphorylation coupled to the oxidation of ferrocytochrome c. J. of Biol. Chem. **215**, 555 (1955).
Nisman, B.: The Stickland reaction. Bacter. Rev. **18**, 16 (1954).
Ochoa, S., A. H. Mehler and A. Kornberg: Biosynthesis of dicarboxylic acids by carbon dioxide fixation. I. Isolation and properties of an enzyme from pigeon liver catalyzing the reversible oxidative decarboxylation of l-malic acid. J. of Biol. Chem. **174**, 979 (1948).
Oginsky, E. L., and R. F. Gehrig: The arginine dihydrolase system of Streptococcus faecalis. III. The decomposition of citrulline. J. of Biol. Chem. **204**, 721 (1953).

OKA, Y.: Heat of formation of metabolic substances. I. Heat of formation of l (+) glutamic acid, 1 (−) tyrosine. Nippon Seirigaku Zasshi **9**, 365 (1944)(a). Zit. in Chem. Abst. **41**, 4701 (1947).
— Calculation of free energy by calorimetric determination. I. Specific heat, entropy and free energy of l (+) glutamic acid at low temperature. Nippon Seirigaku Zasshi **9**, 359 (1944)(b). Zit. in Chem. Abst. **41**, 4700 (1947).
OPARIN, A. I.: Orgin of life. New York: The MacMillan Company 1938, 2nd edit. New York: Dover Publications, Inc. 1953.
PARKS, G. S., and H. M. HUFFMAN: The free energies of some organic compounds. New York: Reinhold 1932.
— K. K. KELLEY and H. M. HUFFMAN: Thermal data on organic compounds. V. A revision of the entropies and free energies of nineteen organic compounds. J. Amer. Chem. Soc. **51**, 1969 (1926).
— T. J. WEST, B. F. NAYLOR, P. S. FUJII and L. A. McCLAINE: Thermal data on organic compounds. XXIII. Modern combustion data for fourteen hydrocarbons and five polyhydroxy alcohols. J. Amer. Chem. Soc. **68**, 2524 (1946).
PAUL, K. G.: Oxidation-reduction potential of cytochrome c. Arch. of Biochem. **12**, 441 (1947).
PINCHOT, G. B.: Phosphorylation coupled to electron transport in cell-free extracts of Alcaligenes faecalis. J. of Biol. Chem. **205**, 65 (1953).
PIRIE, N. W.: On making and recognizing life. New Biology, vol. 16, p. 41. London: Penguin Books 1954.
POGELL, P. M., and R. W. McGILVERY: Partial purification of fructose 1:6-diphosphatase. J. of Biol. Chem. **208**, 149 (1954).
PORTER, J. R.: Bacterial chemistry and physiology. London: Chapman & Hall 1946.
QUAYLE, J. R., R. C. FULLER, A. A. BENSON and M. CALVIN: Enzymatic carboxylation of ribulose diphosphate. J. Amer. Chem. Soc. **76**, 3610 (1954).
RABINOVITZ, M., M. P. STULBERG and P. D. BOYER: The control of pyruvate oxidation in a cell-free rat heart preparation by phosphate acceptors. Science (Lancaster, Pa.) **114**, 641 (1951).
RACKER, E.: Crystalline alcohol dehydrogenase from bakers' yeast. J. of Biol. Chem. **184**, 313 (1950).
— Alternate pathways of glucose and fructose metabolism. Adv. Enzymol. **15**, 141 (1954)(a).
— Formation of acyl and carbonyl complexes associated with electron transport and group-transfer reactions. In: A Symposium on the Mechanism of Enzyme Action. Edit. W. D. McELROY and B. GLASS. Baltimore: Johns Hopkins Press 1954(b).
— Personal communication 1954(c).
— Synthesis of carbohydrates from carbon dioxide and hydrogen in a cell-free system. Nature (Lond.) **175**, 249 (1955).
— G. L. DE LA HABA and I. G. LEDER: Thiamine pyrophosphate, a coenzyme of transketolase. J. Amer. Chem. Soc. **75**, 1010 (1953).
—, and I. KRIMSKY: Glutathione, a prosthetic group of glyceraldehyde-3-phosphate dehydrogenase. J. of Biol. Chem. **198**, 721, 731 (1952).
RECKNAGEL, R. O., and V. R. POTTER: Mechanism of the ketogenic effect of ammonium chloride. J. of Biol. Chem. **191**, 263 (1951).
REED, L. J.: Metabolic functions of thiamine and lipoic acid. Physiologic. Rev. **33**, 544 (1953).
— B. G. DE BUSK, I. C. GUNSALUS and C. S. HORNBERGER: Crystalline α-lipoic acid: A catalytic agent associated with pyruvate dehydrogenase. Science (Lancaster, Pa.) **114**, 93 (1951).
ROSENBLUETH, A., N. WIENER and J. BIGELOW: Behaviour, purpose and teleology. Philosophy of Science **10**, 18 (1943).
ROSENFELD, B., and E. SIMON: The mechanism of the butanol-acetone fermentation. I. The role of pyruvate as an intermediate. J. of Biol. Chem. **186**, 395 (1950)(a).

Rosenfeld, B., and E. Simon: The mechanism of the butanol-acetone fermentation. II. Phosphoenolpyruvate as a new intermediate. J. of Biol. Chem. **186**, 405 (1950) (b).

Rossini, F. D.: Heats of combustion and of formation of the normal aliphatic alcohols in the gaseous and liquid states and the energies of their alcoholic linkages. J. Res. Nat. Bur. Stand. **13**, 189 (1934).

— D. D. Wagman, W. H. Evans, S. Levine and I. Jaffe: Selected value of chemical thermodynamic properties. Nat. Bur. Stand., Circular **1952**, No 500.

Sanadi, D. R., D. M. Gibson and P. Ayengar: Guanosine triphosphate, the primary product of phosphorylation coupled to the breakdown of succinyl coenzyme A. Biochim. et Biophysica Acta **14**, 434 (1954).

— — — and L. Ouellet: Evidence for a new intermediate in the phosphorylation coupled to α-ketoglutarate oxidation. Biochim. et Biophysica Acta **13**, 146 (1954).

— J. W. Littlefield and R. M. Bock: Studies on α-ketoglutaric oxidase. II. Purification and properties. J. of Biol. Chem. **197**, 851 (1952).

Santer, M., and W. Vishniac: CO_2 incorporation by extracts of Thiobacillus thioparus. Biochim. et Biophysica Acta **18**, 157 (1955).

Scheffer, M. A.: De suikervargisting door bacterien der coli-groep. Thesis Delft 1928.

Schepartz, B.: Inhibition and activation of the oxidation of homogentisic acid. J. of Biol. Chem. **205**, 185 (1953).

—, and S. Gurin: The intermediary metabolism of phenylalanine labelled with radioactive carbon. J. of Biol. Chem. **180**, 663 (1949).

Schmidt, G. C., M. A. Logan and A. A. Tytell: The degradation of arginine by Clostridium perfringens (BP6K). J. of Biol. Chem. **198**, 771 (1952).

Scott, D. B. M., and S. S. Cohen: Enzymatic formation of pentose phosphate from 6-phosphogluconate. J. of Biol. Chem. **188**, 509 (1951).

— — The oxidative pathway of carbohydrate metabolism in Escherichia coli. I. The isolation and properties of glucose 6-phosphate dehydrogenase and 6-phosphogluconate dehydrogenase. Biochemic. J. **55**, 23 (1953).

Seubert, W., and F. Lynen: Enzymes of fatty acid cycle. II. Ethylene reductase. J. Amer. Chem. Soc. **75**, 2787 (1953).

Shaw, D. R. D.: Polyol dehydrogenases: galactitol and D-iditol dehydrogenases. Ph. D. Thesis University of New Zealand 1956.

Simon, E.: The formation of lactic acid by Clostridium acetobutylicum (Weizmann). Arch. of Biochem. **13**, 237 (1947).

Skinner, H. A.: Thermochemistry. Ann. Rep. Chem. Soc. **51**, 33 (1954).

Slade, H. D., and W. C. Slamp: The formation of arginine dihydrolase by Streptococci and some properties of the enzyme system. J. Bacter. **64**, 455 (1952).

—, and C. W. Werkman: The anaerobic dissimilation of citric acid by cell suspensions of Streptococcus paracitrovorus. J. Bacter. **41**, 675 (1941).

Slater, E. C.: A comparative study of the succinic dehydrogenase cytochrome system in heart muscle and in kidney; the action of inhibitors on the systems of enzymes which catalyse the aerobic oxidation of succinate; a respiratory catalyst required for the reduction of cytochrome c by cytochrome b. Biochemic. J. **45**, 1, 8, 14 (1949).

— Respiratory chain phosphorylation. Proc. 3rd. Int. Congr. of Biochem., Brussels 1955, p. 264, Edit. by C. Liebecq. New York: Academic Press 1956.

Slein, M. W.: Phosphomannose isomerase. J. of Biol. Chem. **186**, 753 (1950).

Sprinson, D. B.: The biosynthesis of shikimic acid from labelled carbohydrates. In: A Symposium on Amino Acid Metabolism, p. 817, edit. by W. D. McElroy and B. Glass. Baltimore: Johns Hopkins Press 1955.

Srere, P. A., J. R. Cooper, V. Klybas and E. Racker: Xylulose 5-phosphate, a new intermediate in the pentose phosphate cycle. Arch. of Biochem. a. Biophysics **59**, 535 (1955).

— H. L. Kornberg and E. Racker: Conversion of pentose phosphate to hexose phosphate catalyzed by purified enzymes. Federat. Proc. **14**, 285 (1955).

Srinivasan, P. R., M. Katagiri and D. B. Sprinson: The enzymatic synthesis of shikimic acid from D-erythrose-4-phosphate and phosphoenolpyruvate. J. Amer. Chem. Soc. **77**, 4943 (1955).

Stadtman, E. R.: The net enzymic synthesis of acetyl coenzyme A. J. of Biol. Chem. **196**, 535 (1952).

—, and H. A. Barker: Fatty acid synthesis by enzyme preparations of Clostridium kluyveri. J. of Biol. Chem. **184**, 769 (1950).

— M. Doudoroff and F. Lipmann: The mechanism of acetoacetate synthesis. J. of Biol. Chem. **191**, 377 (1951).

— G. D. Novelli and F. Lipmann: Coenzyme A function in, and acetyl transfer by, the phosphotransacetylase system. J. of Biol. Chem. **191**, 365 (1951).

Stanier, R. Y.: Some singular features of bacteria as dynamic systems. p. 3. In: Cellular Metabolism and Infections. Edit. E. Racker. New York: Academic Press 1954.

Stadie, W. C.: Current concepts of the actions of insulin. Physiologic Rev. **34**, 52 (1954).

Stephenson, M.: Bacterial Metabolism. London: Longmans, Green & Co. 1949.

Stern, J. R., and A. del Campillo: Enzymic reaction of crotonyl coenzyme A. J. Amer. Chem. Soc. **75**, 2277 (1953).

—, and S. Ochoa: Enzymatic synthesis of citric acid. I. Synthesis with soluble enzymes. J. of Biol. Chem. **191**, 161 (1951).

— — and F. Lynen: Enzymatic synthesis of citric acid. V. Reaction of acetyl coenzyme A. J. of Biol. Chem. **198**, 313 (1952).

Stetten, M. R.: Metabolic relationship between glutamic acid, proline, hydroxyproline and ornithine. In: Symposium on Amino Acid Metabolism, p. 277, edit. by W. D. McElroy and B. Glass. Baltimore: Johns Hopkins Press 1955.

Stiehler, R. D., and H. M. Huffman: Thermal data. V. The heat capacities, entropies and free energies of adenine, hypoxanthine, guanine, xanthine, uric acid, allantoin and alloxan. J. Amer. Chem. Soc. **57**, 1741 (1935).

Stokes, J. L.: Fermentation of glucose by suspensions of Escherichia coli J. Bact. **57**, 147 (1949).

Strecker, H. J., and I. Harary: Bacterial butylene glycol dehydrogenase and diacetyl reductase. J. of Biol. Chem. **211**, 263 (1954).

—, and S. Korkes: Glucose dehydrogenase. J. of Biol. Chem. **196**, 769 (1952).

—, and P. Mela: The interconversion of glutamic acid and proline. Biochem. et Biophysica Acta **17**, 580 (1955).

Strittmatter, C. F., and E. G. Ball: A hemochromogen component of liver microsomes. Proc. Nat. Acad. Sci. U.S.A. **38**, 19 (1952).

Suda, M., and Y. Takeda: Metabolism of tyrosine. II. Homogentisicase. J. of Biochem. (Tokyo) **37**, 381 (1950).

Swanson, M. A.: Phosphatases of liver. 1. Glucose 6-phosphatase. J. of Biol. Chem. **184**, 647 (1950).

Talalay, P., and M. M. Dobson: Purification and properties of β-hydroxy and steroid dehydrogenase. J. of Biol. Chem. **205**, 823 (1953).

—, and P. I. Marcus: Enzymic formation of 3-α-hydroxy steroids. Nature (Lond.) **173**, 1189 (1954).

Talley, E. A., and A. S. Hunter: Solubility of lactose and its hydrolytic properties. J. Amer. Chem. Soc. **74**, 2789 (1952).

Tatum, E. L., S. R. Gross, G. Ehrensvärd and L. Garnjobst: Synthesis of aromatic compounds by Neurospora. Proc. Nat. Acad. Sci. U.S.A. **40**, 271 (1954).

Taylor, J. B., and J. S. Rowlinson: The thermodynamic properties of aqueous solutions of glucose. Trans. Faraday Soc. **51**, 1183 (1955).

Terrell, A. W.: Succinic acid and glucose in pituitary ketonuria. Proc. Soc. Exper. Biol. a. Med. **39**, 300 (1938).

Tietz, A., and B. Shapiro: The synthesis of glycerides in liver homogenates. Biochim. et Biophysica Acta **19**, 374 (1956).

TISSIERES, A., and E. C. SLATER: Respiratory chain phosphorylation in extracts of Azotobacter vinelandii. Nature (Lond.) **176**, 736 (1955).

TOPPER, Y. J., and A. B. HASTINGS: A study of the chemical origins of glycogen by use of C^{14} labelled carbon dioxide, acetate and pyruvate. J. of Biol. Chem. **179**, 1255 (1949).

TRUDINGER, P. A.: Phosphoglycerate formation from pentose phosphate by extracts of Thiobacillus denitrificans. Biochim. et Biophysica Acta **18**, 581 (1955).

UDENFRIEND, S., and C. MITOMA: Conversion of phenylalanine to tyrosine. In: A Symposium on Amino Acid Metabolism, p. 876, edit. by W. D. McELROY and B. GLASS. Baltimore: Johns Hopkins Press 1955.

UREY, H. C.: The planets. New Haven: Yale University Press 1952.

UTTER, M. F., and K. KURAHASHI: Mechanisms of action of oxaloacetic decarboxylase from liver. J. of Biol. Chem. **188**, 847 (1953).

—, and H. G. WOOD: Mechanisms of fixation of carbon dioxide by heterotrophs and autotrophs. Adv. Enzymol. **12**, 41 (1951).

VELICK, S. F.: The alcohol and glyceraldehyde 3-phosphate dehydrogenases of yeast and mammals. In: A Symposium on the Mechanism of Enzyme Action, edit. by W. D. McELROY and B. GLASS. Baltimore: Johns Hopkins Press 1954.

VESTLING, C. S.: Standard potential of the old yellow enzyme of yeast. Federat. Proc. **14**, 297 (1955).

VOGEL, H. J.: On the glutamate-proline-ornithine interrelation in various micro-organisms In: A Symposium on Amino Acid Metabolism, p. 335, edit. by W. D. McELROY and B. GLASS. Baltimore: Johns Hopkins Press 1955.

WAINIO, W. W.: Reduction of cytochrome oxidase with ferrocytochrome c. J. of Biol. Chem. **216**, 593 (1955).

WARBURG, O., and W. CHRISTIAN: Verbrennung von Robison-Ester durch Triphospho-Pyridin-Nucleotid. Biochem. Z. **287**, 440 (1936).

— — Abbau von Robisonester durch Triphospho-Pyridin-Nucleotid. Biochem. Z. **292**, 287 (1937).

— — Isolierung und Kristallisation des Proteins des oxydierenden Gärungsferments. Biochem. Z. **303**, 40 (1939).

— — and A. GRIESE: Wasserstoffübertragendes Co-Ferment, seine Zusammensetzung und Wirkungsweise. Biochem. Z. **282**, 157 (1935).

WEBER, H. H.: Adenosine triphosphate and mobility of systems. Harvey Lect. **49**, 37 (1954).

— Das kontraktile System von Muskel und Zellen. Proc. 3rd. Int. Congr. of Biochem., Brussels 1955, p. 356, edit. by C. LIEBECQ. New York: Academic Press 1956.

—, and H. PORTZEHL: The transference of the muscle energy in the contraction cycle. In: Progress in Biophysics, vol. 4, p. 60, edit. by J. A. V. BUTLER and J. T. RANDALL. New York: Academic Press 1954.

WEIL-MALHERBE, A., and A. D. BONE: The hexokinase activity of rat-brain extracts. Biochimic. J. **49**, 339 (1951).

WEINHOUSE, S., and R. H. MILLINGTON: Ketone body formation from tyrosine. J. of Biol. Chem. **175**, 995 (1948).

— — Ketone body formation from tyrosine. J. of Biol. Chem. **181**, 645 (1949).

WEISSBACH, A., and B. L. HORECKER: Enzymatic formation of phosphoglyceric acid from ribulose diphosphate and CO_2. Federat. Proc. **14**, 302 (1955).

— B. L. HORECKER and J. HURWITZ: The enzymatic formation of phosphoglyceric acid from ribulose diphosphate and carbon dioxide. J. of Biol. Chem. **218**, 795 (1956).

— P. Z. SMYRNIOTIS and B. L. HORECKER: Pentose phosphate and CO_2 fixation with spinach extracts. J. Amer. Chem. Soc. **76**, 3611, (1954) (a).

— — — The enzymatic formation of ribulose diphosphate. J. Amer. Chem. Soc. **76**, 5572 (1954) (b).

WOLFE, R. S., and D. J. O'KANE: Cofactors of the carbon dioxide exchange reaction of Clostridium butyricum. J. of Biol. Chem. **215**, 637 (1955).

WOLFROM, M. L., and W. L. LEWIS: The reactivity of the methylated sugars. II. The action of dilute alkali on tetramethyl glucose. J. Amer. Chem. Soc. **50**, 837 (1928).

WOOD, H. G.: The fixation of carbon dioxide and the interrelationships of the tricarboxylic acid cycle. Physiologic. Rev. **26**, 198 (1946).

— Significance of alternate pathways in the metabolism of glucose. Physiologic. Rev. **35**, 841 (1955).

— R. W. BROWN and C. H. WERKMAN: Mechanism. of the butyl alcohol fermentation with heavy carbon acetic and butyric acids and acetone. Arch. of Biochem. **6**, 243 (1945).

— R. W. STONE and C. R. WERKMAN: The intermediate metabolism of the propionic acid bacteria. Biochemic. J. **31**, 349 (1937).

— and C. H. WERKMAN: Pyruvic acid dissimilation of glucose by the propionic acid bacteria. Biochemic. J. **28**, 745 (1934).

WOODS, D. D.: Hydrogenase. IV. The synthesis of formic acid by bacteria. Biochemic. J. **30**, 515 (1936).

WURMSER, R., et S. FILITTI-WURMSER: Sur l'équilibre entre l'alcool isopropylique et l'acétone en solution en présence d'alcooldeshydrase. Potential d'oxydo-réduction du système — CHOH $\rightleftarrows$ CO. J. Chim. physique **33**, 577 (1936).

YANIV, H., and C. GILVARG: Quoted by B. D. Davis 1955.

ZELITCH, I.: The isolation and action of crystalline glyoxylic acid reductase from tobacco leaves. J. of Biol. Chem. **216**, 553 (1955).